THE MASTERY OF MOVEMENT

The Mastery of Movement

Rudolf Laban

FOURTH EDITION
Revised and Enlarged

by
Lisa Ullmann

Formerly Director of the Laban Art of Movement Centre

Dance Books • Alton

First published in 1950 under the title
The Mastery of Movement on the Stage

This revised fourth edition was first published
in 1980, and republished in 2011 by
Dance Books Ltd., Alton, Hampshire, UK

ISBN 978-185273145-8

Preface to the First Edition

The reader may be acquainted with the well-known Chinese story of the centipede which, becoming immobilised, died of starvation because it was ordered always to move first with its seventy-eighth foot, and then to use its other legs in a particular numerical order. This story is often quoted as a warning against the presumption of attempting a rational explanation of movement. But, clearly, the unfortunate insect was the victim of purely mechanical regulations, and that has little to do with the free-flowing art of movement.

The source whence perfection and final mastery of movement must flow is the understanding of that part of the inner life of man where movement and action originate. Such an understanding furthers the spontaneous flow of movement, and guarantees effective liveliness. Man's inner urge to movement has to be assimilated to the acquisition of external skill in movement.

There exists an almost mathematical relationship between the inner motivation of movement and the functions of the body; and guidance in the knowledge of application of the common principles of impulse and function is the only means that can promote the freedom and spontaneity of the moving person.

It is, of course, a difficult task to guide anyone to the mastery of practical activity by means of the printed page, and especially is this the case when the attempt to do so is concerned with the subject we are now considering. To deal with the stage — the mirror of man's physical, mental, and spiritual existence — raises many problems. No doubt we have here some of the reasons why the treatment of our subject has been so rarely undertaken. But a renewed attempt has to be made, and it is hoped that the present treatise will prove an incentive to further research in so important a subject.

The reader should not expect to find in this book a fine literary, or even an easily readable, exposition of the problems involved, for movement has very little to do with literature. The art of movement is an almost self-contained discipline which speaks for itself, and mostly in its own special idiom. Almost every sentence in this exposition is written as an incentive to personal mobility, so the reader should be prepared to avail himself or herself to these incentives. It is hoped that the perusal of the text itself will indicate how to accomplish what is really a kind of mobile reading. Those who prefer to remain comfortable in their chairs while they read will have to skip certain sections of the book, yet there is consolation even for them; for the study of this most interesting subject, viz., *thinking* in terms of movement, is one of our purposes. To assume that mobile thinking merely implies cavorting in the world of ideas is as great an error as to assume that the art of movement on the stage is restricted solely to ballet. Movement is equally an essential means of artistic expression in drama and in opera; and it is also the means of satisfaction and comfort in situations of work, since movement, when scientifically determined, forms the common denominator to both art and industry.

To the examples given in our later chapters we have added a selection of mime scenes. In trying to perform parts of these scenes the reader will be able to put into practice his growing knowledge of the art of movement.

This book embodies the practical studies and experience of a lifetime, but I could not have written it without close exchange of opinions with my friends and pupils. It is therefore a digest of my lectures and talks with very many of my co-workers, some of whom are of international reputation. My thanks are therefore due to all those who have shared my work on the stage and my researches into the art of movement. They are too numerous for individual mention, and to single out a few of them would be invidious, since all in their several degrees have ardently striven to revive and add to the ancient knowledge of the art of movement, and to apply this knowledge to practical life. But all these my coadjutors were present with me in thought as I wrote, and so I gratefully dedicate what I have written to them all

In this guide to stage (and incidentally to factory) practice I have been obliged to work on my own special pattern. Why this was necessary, study of the text will disclose.

R. LABAN

Manchester, January 18, 1950

Preface to the Second Edition

In comparison with the wealth of evidence that Rudolf Laban collected during his lifetime about human movement, and considering his profound comprehension of its significance in individual and communal living, he published little about his findings. He preferred to lead people personally in the world of movement, and he was a brilliant guide, for he had also a deep understanding of people and a feel for their needs. Now, since we are no longer able to experience Laban's personal guidance, his books as well as his many notes and manuscripts are the only source of further inspiration and information about his thoughts and discoveries.

The first edition of this book went out of print shortly before Laban's death in July 1958. He had intended revising the text for the reprint and had discussed with me what improvements he would like to make. It was, however, not granted to him to undertake any changes himself; the task fell upon me. I am aware of the great responsibility which I have assumed in attempting to revise this book, and I would feel even more uncertain about it if I had not known Laban's plans for the new edition.

I have re-written chapters on "Movement and the Body", although basically following the previous outline. In many instances, however, I have had to broaden the content, particularly in the section of "Correlation of Bodily Actions and Effort". While in the original book Laban only hinted at some of his findings with regard to human movement and effort, and introduced the reader to but a few of his theories in general terms, I have tried to give more specific explanations of these.

This book was originally published as *The Mastery of Movement on the Stage*, but although his references were made to the stage in the theatre, Laban used this much as a forum to present his ideas on movement in relation to the stage of human life. He refers in this book to theatrical acting as being "the artistic enhancement of human action". It will surely become apparent to the reader that this artistic enhancement is nothing other than an augmentation of the art of living by learning to master movement. Because of this I have ventured to omit, in the present title, the adjunct "on the Stage", and hope not only that the subject will have appeal for those concerned with the theatre, but that its treatment will give inspiration also to those who seek to understand movement as a force of life.

The terminology used in this field of study and evolved on the basis of Laban's analysis of movement was developed by him and his co-workers in the course of extensive practical experiments and discussions with many people over many years. While a certain degree of clarification has been reached with growing knowledge and experience of the subject, certain modifications of terminology are to be expected. With this in mind, I have, wherever necessary, added a definition of terms. The opportunity has been taken to improve the typography, as well as the whole presentation of the book, in an endeavour to facilitate its study.

In conclusion, I should like to acknowledge with gratitude the help given by my students, who so cheerfully assisted me in my various practical "try-outs". I am particularly indebted to my friends and colleagues who read the text and made most valuable suggestions and gave me such generous help.

It is my sincere hope that the reader will share with me my desire to inquire further into this fascination phenomenon of movement which has been particularly stimulated in revising this book.

LISA ULLMANN

Addlestone, Surrey
October 1960

PREFACE TO THE THIRD EDITION

It is now eleven years since the text of this book was revised and I welcome the opportunity of a new edition to make some amendments and enlargements. Some paragraphs I have re-introduced from Laban's original version of *The Mastery of Movement on the Stage*, others stem from his personal notes on the subject, and I have re-written parts of Chapter 5. There were also a number of corrections which had to be made. With all these additions I hope to have contributed to the further clarification of the subject, although I am fully aware that much remains to be done. This continues to be a challenge to me.

LISA ULLMANN

Addlestone, Surrey
May 1971

PREFACE TO THE FOURTH EDITION

For a long time I have been thinking that a certain marginal legend might assist the reader in gaining more easily a survey of the many important points made in this book. Therefore, I have now in this fourth edition attempted to indicate the subject matter referred to in a particular section.

Since there is an appreciably growing number of people who read and write movement notation — Kinetography Laban — I have added to most of the examples in Chapters 2 and 3 the appropriate kinetograms as Laban had originally intended.

Lastly I have attached an Appendix with a text drawn mainly from an unpublished book manuscript by Laban on "Effort", written before 1950, which he subtitled "Rhythmic Control of Man's Activities in Work and Play". In doing this I hope to meet the many demands from students of movement for an outline of basic considerations which are relevant to the practice of Effort observation, assessment and notation.

LISA ULLMANN

Addlestone, Surrey
November 1979

Contents

CHAPTER 1

Introduction

Man moves in order to satisfy a need. He aims by his movement at something of value to him. It is easy to perceive the aim of a person's movement if it is directed to some tangible object. Yet there also exist intangible values that inspire movement.

Eve, our first mother, in plucking the apple from the tree of knowledge made a movement dictated both by a tangible and an intangible aim. She desired to possess the apple in order to eat it, but not solely to satisfy her appetite for food. The Tempter told her that by eating the apple she would gain supreme knowledge: that knowledge was the ultimate value she desired.

Can an actress represent Eve plucking an apple from a tree in such a way that a spectator who knows nothing of the biblical story is made aware of both her aims, the tangible and the intangible? Perhaps not convincingly, but the artist playing the role of Eve can pluck the apple in more than one way, with movements of varying expression. She can pluck the apple greedily and rapidly or languidly and sensuously. She can, too, pluck it with a detached expression in the outstretched arm and grasping hand, in her face and in her body. Many other forms of action are possible, and each of these will be characterised by a different kind of movement.

In defining the kind of movement as greedy, as sensuous, or detached, one does not define merely what one has actually seen. What the spectator has seen may have been only a peculiar, quick jerk or a slow gliding of the arm. The impression of greed or sensuousness is the spectator's personal interpretation of Eve's state of mind in a definite situation. If he should observe Eve grasping quickly into the air, that is to say, if he sees the same movement performed without the objective aim, he would probably not be induced to think either of the object or of the motive. He would perceive the quick, grasping movement without understanding its dramatic significance.

1

*Momentary moods and
constant features of
personality revealed in
movement*

It may, of course, occur to the spectator to ask whether this movement, apparently without an objective aim, was made in order to reveal certain traits of Eve's mood, or of her character. It is unlikely that one movement could convey to him more than a superficial impression, since it could never give a definite picture of her character. On the other hand, several concurrent movements, as, for example, those including Eve's carriage and walking before the snatching gesture, would offer additional and clearer indications of her personality. But even then Eve's behaviour in the act of plucking the apple would be less characteristic of her personality than of her momentary eagerness in the particular situation. On other occasions she might develop quite different rhythms and shapes of movement which, comprising several actions, would show her as a person in an altogether different light.

So movement evidently reveals many different things. It is the result of the striving after an object deemed valuable, or of a state of mind. Its shapes and rhythms show the moving person's attitude in a particular situation. It can characterise momentary mood and reaction as well as constant features of personality. Movement may be influenced by the environment of the mover. So, for instance, the milieu in which action takes place will colour the movements of an actor or of an actress. They will be different in the role of Eve in paradise, or of a society woman in an eighteenth-century salon, or of a girl in the bar of a public-house in the slums. All three women might be similar personalities exhibiting almost the same general movement characteristics, but they would adapt their behaviour to the atmosphere of an epoch, or a locality.

A character, an atmosphere, a state of mind, or a situation cannot be effectively shown on the stage without movement, and its inherent expressiveness. Movements of the body, including the movements of the voice-producing organs, are indispensable to presentation on the stage.

*People move in order to
meet and part*

There is yet another aspect of movement which is of paramount importance in acting. When two or more players are to meet on the stage, they have to make their entrance, their approach to one another (by either touching, or keeping a due distance), and later they have to separate and make their exit.

The grouping of the actors on the stage occurs in movement, which is expressive in another sense than individual movement. The members of a group move in order to show their desire to get in touch with one another. The ostensible object of meeting might be to fight or to embrace, or to dance, or just to converse. But there exist intangible objectives, such as the attraction between sympathetic individuals or the repulsion felt by persons or groups antipathetic to one another.

Group movement

Group movements can be brisk and pregnant with the threat of aggression, or soft and sinuous, like the movement of water in a placid lake. People can group themselves as hard, detached rocks on a mountain or as on a leisurely flowing stream in a plain. Clouds frequently form most interesting groupings which produce a strangely dramatic effect. Group movements on the stage resemble in a way the shifting clouds from which either thunder rolls, or sunshine breaks.

The individual actor will sometimes use his movements as if his limbs were the members of a group, and this is probably the solution of the riddle of the expressiveness of gesture. When Eve plucks the apple in a greedy or a languid manner, she will express her attitude by the movement of parts of her body. In greed her arms, or even her whole body, may shoot out suddenly, avidly, all in one and the same direction towards the coveted object. The languid approach may be characterised by a nonchalant, slow lifting of an arm, while the rest of the body is lazily curved away from the object. It is almost a dancelike movement, in which the outward action is subordinate to the inner feeling. No words are needed to convey this feeling to the spectator.

The riddle of expressiveness of gesture

Eve reaching for the apple is not yet a dramatic scene. Drama starts when she offers the apple to Adam. Drama always happens between two or more people, and here is where group movement comes into its own. Even a soliloquy and a solo dance are in reality a dialogue between the two poles of an individuality moved by personal reflections or changing moods. The duality of the poles becomes visible in movements which display the inner tensions.

When drama starts

In scenes of love and combat, the duality of the emotions is embodied in two real persons. Eve offers Adam the forbidden fruit. Her gesture of offering and his of accepting are more than a utilitarian move in which the apple passes from one to the other. It is a gathering storm, presaging thunder-clouds heavy with fate of the race. The gestures will be less conspicuous and less expressive if the impending thunderstorm is condensed into a spoken dialogue. The story of the forbidden fruit represented in a dance-mime will be enlivened more by elaborate gestures than the same story accompanied by words.

Pure dancing has no describable story. It is frequently impossible to outline the content of a dance in words, although one can always describe the movement. In pure dance the spectator could not know that a movement of grasping quickly into the air would express greed or any other emotion in relation to the apple. He would only see the quick grasp and experience the meaning of it through the play of rhythms and shapes which in dance tells its own story, a frequent happening in the world of not logically defined values and longings.

Pure dancing and dance-mime

Movement in pure dancing does not need to adapt itself to characters, actions, epochs and situations, but it does in dance-mime, which is virtually acting without words, although often supported by background music. The performance of social dances on the stage characteristic of a historical period, of the social status of the people, and of the occasion and locality of the dance, cannot be considered as pure dancing. In pure dancing the inner drive to move creates its own patterns of style, and of striving after intangible and mostly indescribable values.

The art of movement embraces the whole range of bodily expression

The art of movement on the stage embraces the whole range of bodily expression including speaking, acting, miming, dancing and even musical accompaniment.

Movement used to attain tangible and intangible values; work and prayer

Spoken drama and musical dance are, however, late flowers of human civilisation. Movement has always been used for two distinct aims: the attainment of tangible values in all kinds of work, and the approach to intangible values in prayer and worship. The same bodily movements occur in both work and worship, but their significance differs. In order to achieve the practical purpose of work, the stretching out of an arm and the gripping and handling of an object have to be made in a logical order. Not so in worship. Here gestures follow one another in an entirely irrational sequence, though each of the gestures used in worship might also be part of a working action. The stretching out of an arm in the air might express the longing for something which cannot be grasped. The swinging of the arms and the body, which might resemble the handling of an object, might signify an inner struggle and become the expression of a prayer for liberation from inner turmoil.

Development of human expressiveness

The European has lost the habit and capacity to pray with movement. The vestiges of such praying are the genuflexions of the worshippers in our churches. The ritual movements of other races are much richer in range and expressiveness. Late civilisations have resorted to spoken prayer in which the movements of the voice organs become more important than bodily movements. Speaking is then often heightened into singing.

It is, however, probable that liturgical praying and ritual dancing co-existed in very early times; and so it is also probable that the spoken drama and the musical dance have both developed from worship; from liturgy on the one hand and ritual on the other. The whole complexity of human expressiveness, as comprised in the art of movement, is represented in the above diagram.

We never know whether man regards himself as taking part in a tragedy or a comedy with himself as the protagonist in the drama of existence and Nature forming the chorus. Yet it is an undeniable fact that man's extraordinary power of thought and action has placed him in a peculiar situation so far as his relationship to his surroundings is concerned. *Man's peculiar situation in respect of his surroundings*

Man tries to enact the conflicts arising from the solitary role of his race. Reflected in the mirror of tragi-comedy the public sees a character struggling and falling either to destruction or into ridicule. The tears for the first, as well as the laughter for the second, with which the public reacts to the representation of the character's inner and outer adventures appear to be equally comforting.

This is not a utilitarian explanation of, or an excuse for, acting. There is more beneath the co-operation of audience and actors than the fun of contemplating man's misery or folly, one eye laughing and the other weeping. The theatre mirrors more than the everyday world of our sufferances and gaieties. The theatre gives an insight into the workshop in which man's power of reflection and action is generated. This insight offers more than an increased understanding of life; it offers the inspiring experience of a reality transcending that of our everyday fears and satisfactions. *The theatre offers more than an increased understanding of life*

What really happens in the theatre does not occur only on the stage or in the audience, but within the magnetic current between both these poles. The actors on the stage forming the active pole of this magnetic circuit are responsible for the integrity of purpose with which a play is performed. On this depends the quality of the exciting current between stage and audience.

The charm of the mechanical perfection of the movements of speech and gesture is undeniable. It is an enormous pleasure to hear well delivered speech, and to witness appropriate gestures in which the victory of mind over matter appears complete.

The individual actor who achieves this charm of presentation stands on an elevated rung of the ladder of perfection. But a question arises: is it the highest rung? Virtuosity of this kind uses the movements of the body and also of the voice-producing organs, as the skilled labourer uses his tools. The high economy of effort which characterises skill is common to both *Skill and economy of effort on the stage and in the workshop* the labourer and the virtuoso. The greater the economy of effort the less apparent is the strain. High economy of effort makes movement look almost effortless. This applies also to voice-producing movements, heard but not seen. Expressive details remain almost accidental in this kind of implemented movement, because the whole effort is concentrated upon the smooth performance of the actions necessary to the work. The labourer's job is to work with material objects, the actor's to work with his own body and voice and to employ them in such a way that human personality and its changing behaviour in various situations are effectively characterised. Both the labourer's and the actor's job can be done skilfully with a pleasant and useful economy of effort. But in the actor's case more is required. He has to communicate with his audience. He has to establish that contact between stage and audience that has been compared above to a bi-polar magnetic current. It is obvious that the exhibition of skill in a particular movement can create some kind of contact, sometimes even a most intimate and satisfying contact.

Here the question of the quality of the contact arises. The integrity of purpose we have postulated might be considered a question of taste. The artist who prefers to use skilled implemental movement certainly stands high on the ladder of perfection. In his presentation the theatre is pure entertainment, mirroring man's happiness, folly and misery, in which the audience can find comfort and relief from its workaday sorrows.

Let us suppose the contrary to be true. The conclusions then to be drawn are interesting, and are supported by the example of many first-class artists who are not virtuosos. The actor who tries to do more than represent life in a skilful manner uses the movements of his body and his voice-producing organs with his interest focused on that which he intends to convey to his audience and less on the external shapes and rhythms of his actions. This kind of performer concentrates on the actuation of the inner springs of conduct preceding his movements, and pays little attention at first to the skill needed for presentation. A different quality of contact with the public thus results if, instead of skill, the inner participation is stressed.

The role of skill and the actuation of inner springs preceding movement expression

It is obvious that the virtuoso will easily be tempted to restrict the number of his movements to those which best suit his skill. The other type of actor will be inclined to reject all selection and almost any exercise of single movement forms which to him are mere acrobatics. In his endeavour to get his spontaneous movements to flow freely, he will often be more erratic and impulsive than the virtuoso.

On the whole it can be said that these two contrasting viewpoints apply movement to two differing aims: on the one hand, to the representation of the more external features of life, and on the other, to the mirroring of the hidden processes of the inner being.

The actor striving for the second of these two aims has a deeper inclination, and a greater chance to penetrate into the remoter recesses of what we have called the workshop of thought and action. In putting this kind of actor — provided he is perfect in his own type — on a higher rung on the ladder of theatrical values, preference is given to the less primitive, and therefore more complex, form of human mentality and taste. This preference could to a certain extent be justified today, because it seems that modern man needs a deeper penetration into the innermost recesses of life and of human existence, which, if brought to the surface, might help him to recover some lost but essential qualities. People who seem to have grown out of the worship of pure brilliancy of movement seek a new target for their urge to admire. Much depends on the dramatists and choreographers, and the type of play or ballet they present; although it is a fact that actors or dancers possessing the real creative sense of presentation are able to give to a mediocre play that revealing quality which sheds light on the hidden depths of human nature. Such a vital presentation arises almost always from recognising the fact that the visible and audible means of the performer's expression are exclusively movements.

Movements used in works of stagecraft are those of the body, the voice-producing organs, and, one may add, the motions performed by the instrumentalists of the orchestra. Human movement, with all its physical, emotional, and mental implications, is the common denominator of the dynamic art of the theatre. Ideas and sentiments are expressed by the flow of movement, and become visible in gestures, or audible in music and words. The art of the theatre is dynamic, because each phase of the *The art of the theatre* performance fades away almost immediately after it has appeared. *is dynamic* Nothing remains static, and the leisurely inspection of details is impossible. In music one sound succeeds another, and the first dies before the next is heard. The actors' lines and the dancers' movements are all in a continuous dynamical flux which is interrupted only by short pauses until it finally ceases at the end of the performance.

In theatrical acting, which is artistic enhancement of human action, the controversial tendencies contained in the thinking and feeling of the various characters represented are expressed both in words and in gesture. In mime and ballet the dynamics of thought and emotion are expressed in a purely visible form. They are, as it were, written into the air by the movements of the performer's body. What music, the audible part of a ballet performance, does for dancing is partly to translate their emotional content into sound-waves. In opera, speaking is replaced by singing, whereby the prominent role is given to music. Movement, with its wide *Movement is the common* range of visible and audible manifestations, offers not only a common *denominator in acting,* denominator for all stage work but it also secures the basis for the *miming, dancing, singing,* common animation of all those participating in this creation. *speaking, music and* *sound making*

The fluid transiency of works of the dynamic arts has to be contrasted

The intrinsic nature of the static arts

with the solid durability of works of the static and plastic arts — architecture, sculpture and painting — though one must remember that the last-mentioned contributes to the stage decor and costume. But the dynamics of action and dance are easily killed if the static components of the performance, costume, and decor are over-stressed. The highly pictorial theatre neglects the essential feature of stagecraft, which is movement. This is better understood if one considers the intrinsic nature of static art.

Several objects grouped together may, by their shape, colour and arrangement, awaken in a painter the desire to create a still-life. The picture may outlive the subject and its creator for centuries. Such a work of art, in which the artist has translated his idea into a static, and therefore permanent, form, preserves what would otherwise have been a transient impression. In the same way architecture may outlive its designer. The architect has a sudden vision, an intuitional idea, and he embodies it enduringly in stone. Again, a statue might stand, say, in a garden, and although both the sculptor and his model have passed away centuries ago, the statue remains a witness of the beauty that once inspired the sculptor's chisel.

Artist's dynamic power enshrined in the forms of his work

Static works of art comprise pictures, sculpture, architecturally beautiful buildings, to which we may add utilitarian objects bearing the imprint of the creative impulse of genius. In these art forms the dynamic power of the creator is enshrined in the form of his work. The movements he has used in drawing, painting or modelling have given character to his creations, and they remain fixed in the still visible strokes of his pencil, brush, or chisel. The activity of his mind is revealed in the form he has given to his material.

The static artist creates works which can be seen as a whole and perceived at a glance by looking at the drawings, pictures, sculptures and buildings which his genius has produced. Generations come and go, and each in turn admires the same canvas, vase, statue or building, to which the artist has imparted almost imperishable individual form.

Spectator's response to the impacts of the various theatrical arts

The effect on the onlooker of a work of static art is genuinely different from the effect on the spectator of a theatrical performance. In a picture the mind of the onlooker is invited to go its own way. Memories and the association of ideas conduce to a contemplative mood and meditative inner activity. The audience at a play, a mime or a ballet has no opportunity for contemplation. The spectator's mind is forcibly submerged by the flow of the ever-changing happenings which, given a real inner participation on his part, leaves no time for the elaborate cogitation and meditation, both possible and natural, when viewing, say, a picture, or some scene of natural beauty. Stage performances in which the pictorial element of decor and still-life is over-stressed are apt to impair the spectator's interest in the attention to the dynamic happening which is the all-important element. Not only may such an over-emphasis of pictorial and architectural elements be a danger, but also too much analytical thinking in the speech parts can destroy the essential unity of drama. The intellectual book drama is ineffectual on the stage. Analytical thinking tends to the formation of static ideas and an excess of meditation. When this kind of thinking prevails in a play, the flow characteristic of good dramatic dialogue and controversial action is impeded. Mime, built on movement occurrences both in its content and its form, is the basic theatrical art. Man's conflicting efforts* in his struggle for values are most truly revealed by mime. Too many words and too much music are both apt to overshadow the truth of this effort display as it becomes apparent through the performer's bodily actions.

Mime, the basic theatrical art

Pure mime is now almost unknown, and its loss is of deep concern. The art of mime flourished in the early periods of human civilisation, in humanity's adolescence. There are values, such as youth, ingenuousness and innocence, which can be lost by individuals and races and can never be recovered. It is the same with certain forms of happiness, gaiety, harmlessness and, to some extent, with beauty, charm and gracefulness — all natural gifts of youth and innocence. The struggle for such values, which one cannot attempt to gain or regain, is ridiculous in everyday life.

The miracle of recovering these values, which have been forever lost in ordinary everyday life, is possible on the stage. Why? Because the actor or mime can represent character and circumstance if he knows enough about their inherent effort characteristics. Youth, ingenuousness, harmlessness, beauty, charm, gracefulness are dependent on inner attitudes, and these the actor can consciously reproduce. This seems a paradox, but not so on the stage, where values have not to be possessed, but to be pictured, and this is done by selection and formulation of appropriate effort qualities. Who does this picturing is irrelevant, so long as it is effectively done. One sees middle-aged actresses playing exquisitely the parts of young girls, and able to convey to us the truth about youth and its fate in the most touching manner. Inner youthfulness coincides frequently with what is called virtue; vice is incompatible with innocence; nevertheless the components of the effort qualities displayed by a virtuous and by a vicious person are the same and include the same elements of movement. The arrangement of the inner impulses creating the movement shows, however, differences in rhythm and stress.

In the theatre lost values can be depicted by acting and miming

Sameness of effort components but different meaning through their arrangement

One of the most characteristic primitive dance-mime productions consists of imitations of animal movements. The benevolent or malevolent — virtuous or vicious — spirits governing human fate are represented as

* The inner impulses from which movement originates are in this and other publications of the author called "Effort" (see also page 21).

animals, venerated or abhorred by members of primitive tribes. It is useful for the actor-dancer to consider and to compare the typical movement rhythms of various living beings, animals as well as men, in order to gain insight into the selection of effort qualities, or the kind of inner impulses, appropriate to the various characters, situations and circumstances represented in primitive mime.

Effort manifestations of man and animals

It appears that the effort characteristics of men are much more varied and variable than those of animals. One meets people with cat-like, ferret-like, horse-like movements, but one never sees a horse, a ferret, a cat exhibiting human-like movements. The animal world is rich in effort manifestations, but each animal genus is restricted to a relatively small range of typical qualities. Animals are perfect in the efficient use of the restricted effort habits they possess; man is less efficient in the use of the more numerous effort shadings potentially possible to him. It is not surprising that more numerous and vehement conflicts should arise in human beings possessing the capacity for such manifold and often contradictory combinations of effort qualities.

Movement habits of mammals doubtless approximate closer to those of man than do the movements of animals of lower organisation. Human movement habits are more often compared with those of mammals, and perhaps also of birds, than with those of fishes, reptiles or insects.

A thickly populated street in a big city might be held to suggest a beehive or an anthill, as long as the effort peculiarities of the single individuals in the crowded street are not distinguished. It is not impossible to see in an anthill individual ants with special effort characteristics, some ants being more mobile and energetic than others. In a crowded street, however, one can more easily distinguish individuals who are more mobile and energetic than others, and also those resembling in their effort manifestations, in gait and in gesture, a hundred different animals. When one looks at the facial expressions and movements of people in a crowd one will soon distinguish the personal effort characteristics of the individuals. The crowd reminds one then of a mixed gathering of animals of all kinds. We can no longer compare such a gathering to a swarm of ants because ants and their movements are but little differentiated.

The faces and hands of human beings, and the heads and paws of animals, are often taken as indicative of the similarity between human beings and animals. The faces and hands of human adults could be considered as having been moulded by their effort habits. The shape of their body, including that of the head and the extremities, might signify a natural effort disposition and might be regarded as "frozen" effort

Shapes of constitutional nature and shadow movements

manifestations. However, shapes of a constitutional nature are much less revealing than movements, especially those which might be called shadow movements. These are tiny muscular movements such as the raising of the brow, the jerking of the hand or the tapping of the foot, which have none other than expressive value. They are usually done unconsciously and often accompany movements of purposeful action like a shadow — hence the term. Facial movements often offer great contrasts to face shapes. Handsomely moulded features can become most repellent when grimacing, and ugly people can assume the nicest expressions when smiling. Everyone is familiar with facial expressions that mirror inner conflicts.

Returning now to the simple effort manifestations of animals, it can be said that each order of animals seems to have selected a few of the millions of possible effort combinations, and to have maintained them throughout long generations. These restricted series of effort combinations may have formed the typical body shapes and movement habits of the different species.

A few people display effort qualities typical of animals, but they are not restricted in all their movements to these. On certain occasions, or through training, they can get away from their typical fixation and can easily use effort combinations impossible to the animal they resemble. A person's ability to change the quality of effort, that is, the way in which nervous energy is released, by varying the composition and sequence of its components, together with the reactions of others to these changes, are the very essence of mime. This is that kind of drama in which scenes from life are imitated and it is an activity of which only the human being is capable.

Ability to change the qualities with which nervous energy is released

The movements of a cat or cat-like creature are mostly freely flowing, and this is to the detriment of more restrained movements which are not particularly characteristic of a cat. No one has ever seen a cat strut with a horse-like gait. A cat-like man, however, can sometimes strut like a horse, if he wishes to do so, and he might provoke admiration or ridicule by so doing.

It is not only the fluency which is typical of a cat's movements. When jumping the cat will also be relaxed and flexible. A horse or a deer will bound wonderfully in the air, but its body will be tense and concentrated during the jump. Man's body-mind produces many kinds of qualities. He can jump like a deer, and, if he wishes, like a cat.

The components making up the different effort qualities result from an inner attitude (conscious or unconscious) towards the motion factors of Weight, Space, Time and Flow.

Effort qualities result from an attitude towards W.S.T.F.

The attitude towards these is different with different species. The laziness, that is, an exaggerated indulging in time, of a sloth is proverbial, very much as is the haste, or exaggerated racing against time, of a weasel. Nothing is more tragic than indiscriminate laziness and perpetual haste in man.

The main characteristics of the effort attitudes of a living being are, of course, his more complete effort sequences, and not his attitude towards a single factor of motion. A thorough investigation of the typical sequences of animal effort qualities would be a lifetime's study. The few rudimentary hints given here can only serve the purpose of calling attention to the actual scale of increasing effort complications. Man stands at the top of this scale, because he can use all the shadings of effort an animal can use, and as we shall see later, many more of his own, namely, the special effort combinations, human and humane, which can and do cause more terrific reactions than the simple ones of animals.

Effort capacities of young animals and humans

Another interesting point should be mentioned. Young animals and humans have a much more varied scale of effort capacities at their disposal than their elders. A puppy or a kitten, and in certain respects also a child, is more mobile than a full-grown dog, or cat, or adult human. The typical effort characteristics of the individual and the species are not yet fully developed in the young. The restrictive selection continues after birth, although the full range of typical effort tendencies has been inherited, and is later developed.

The influence of the life story and surroundings during the selection period of early youth becomes clearly visible in certain shades of the final effort characteristics of the grown-up individual. These shades are again much more diversified in man than in animals, excepting in some respects mammals. A domesticated animal, for instance, will develop different shades of effort configuration from a wild animal of the same genus. The necessity for continuous struggle to preserve its existence would make the effort behaviour of a wild animal richer in certain directions, yet more restricted in others, than that of a domesticated animal. Humans are confronted during their period of growth by a struggle for life different from that of the animals. Consequently the development of their effort habits takes a different form. The inherited capacity of an individual to resist retarding influences plays an important part in the final outcome.

Development of effort habits and the possibility of changing them

But humans, whether primitive or civilised, poor or rich, can establish complicated networks of changing effort qualities, representing manifold ways of releasing inherent nervous energy. Man has the capacity to comprehend the nature of the qualities, and to recognise the rhythms and the structures of their sequences. He has the possibility and the advantage of conscious training, which allows him to change and enrich his effort habits even in unfavourable external conditions. Domesticated animals are lost if exposed to the rigours of life incident to free Nature. Fully-grown wild animals can never become entirely domesticated. They have little capacity to change their effort behaviour, but a human being even when he has grown up in primitive surroundings can refine his movement habits if the need arises. Pampered youths can become ferocious men in war, or in other dangerous situations. Such people acquire entirely new effort habits, and they are able in easier situations to return to the former more gentle ones if they so choose. In any case, there is little doubt that man's effort possibilities are both more varied and variable than those of animals, and that this richness is the main source of his dramatic behaviour. A scale could be built up, ranging through many degrees, plotting the most restricted and fixed effort capacities of primitive animals right up to the potentially most complicated and changeable effort attitudes of civilised man.

Besides the comparative richness of human effort capacity, one can notice an effort speciality which might be called the humane effort. No moral evaluation is intended by this statement. Humane effort can be described as effort capable of resisting the influence of inherited or acquired capacities. With his humane effort man is able to control negative habits and to develop qualities and inclinations creditable to man, despite adverse influences. The resulting struggle is full of dramatic implications. *Humane effort*

We are touched by the suggestion of quasi-humane efforts of devotion, sacrifice and renunciation displayed by animals. Such may or may not have a foundation of fact. But we take it for granted that every man is able, and even almost under an obligation, to foster such kinds of effort. Humane effort is rarely taken into full consideration when speaking of movement study and training. But nevertheless it is a most important manifestation and perhaps the very source of the possibility of movement education, which is of paramount importance not only to the actor-dancer but also to every individual's self-development. *Source for making movement education possible*

There is no doubt that the communal instincts of man have developed features that differ greatly from the simple herd instincts of animals. If one investigates the growth of man's communal sense, it will be found that at the basis of historical development there lies a special kind of effort training of which animals are incapable. Those strange effort habits of man, which cannot be entirely explained as adaptation to circumstances and environment, are the result of conscious effort cultivation. Animals drive to sustain their individual and racial life. Man's selection of effort sequences seems no longer to be entirely subconscious; he is able to command a far wider range of effort possibilities than any animal possesses, and this range extends beyond purely life-sustaining necessities. *Life-sustaining effort and beyond*

Man expresses on the stage by carefully chosen effort configurations his inner attitude of mind and he performs a kind of corporate ritual in the presentation of conflicts arising from the differences in these inner attitudes. In domesticating animals, man has learned how to deal with effort and how to change the effort habits of living beings. In applying the principles of domestication to himself he has broadened the scope of effort training into the creation of works of dynamic art. Man can *Scope of effort training* domesticate his fellow men, not just as slaves, but also as happy companions; and he has finally learnt to domesticate himself by training and developing his own personal effort habits, both enlarging their range quantitatively and directing them more and more towards the specific humane forms of effort. The way in which man has achieved this kind of effort education is very remarkable. It has a parallel in the evolution of animal effort habits.

Play of young animals and children

Young animals learn, although not by conscious control, to select and develop their effort qualities in play. Playing animals simulate all kinds of actions which resemble very strongly those real actions they will need to perform to provide for the necessities of their future life. Hunting, fighting, biting seem to be suggested but they do not really hunt, fight and bite, at least not with the aim of procuring food. In young animals and children we call it play; in adult people we call it acting and dancing. During play, effort sequences are tried out, selected and chosen as those best suited, say, for a successful hunt or fight. The young animal, and so also the child, experiments with all imaginable situations: offence, defence, ambush, ruse, flight, fear, but courage is always exhibited. Search for the best possible effort combination for each occasion accompanies these experiments. The body-mind becomes trained to react promptly and with improved effort configurations to all the demands of differing situations until the adoption of the best becomes automatic. The fighting and hunting methods of a cat are not fully developed in a kitten, but the impulse towards them is already discernible in the kitten's first movements. The same is true of a puppy, a lamb, or any other young animal. Play is the great aid to growing effort capacity and effort organisation.

Nothing could prevent our calling these play acrobatics dramatic acting if the words acting and drama were not reserved for man's conscious exhibition of life situations on the stage. There exists also a difference because stage performance demands a spectator whom the actor can address, while the playing puppy, kitten or child is unconcerned

Playing, dancing, acting as effort exercise

with a spectator. The play of young animals is thus nearer to dancing than to acting, since dancing does not always demand spectators. If children and adults dance, that is, perform certain sequences of effort combinations for their own pleasure, no audience is necessary. In this sense dancing is a more genuine effort exercise than acting. Dancing, or at least what we today call dancing, is, nevertheless, different from play, yet it is not in itself a stage art.

The resemblance to life's struggle is not so clearly evident in dancing. Dance is stylised play which is not directly related to dramatic effort behaviour. Some animals dance: that means they have learned to stylise their play in the same way that man has done. If we study the exhaustive descriptions of the movements of birds and apes as observed by scientists, we are amazed by the similarity of these movements to human dancing. While during play the effort qualities of these creatures are intermingled in an almost casual and irregular way, in human dances they are neatly selected, worked out and separated. Regular repetitions of effort sequences form rhythmical phrases and are exactly repeated. What dancing animals intend would remain a riddle if we did not assume that the whole activity takes a form of a more or less conscious effort selection and effort training.

It is a well-known fact that the plays and dances of primitive tribes originate in an endeavour to make themselves aware of certain selected effort combinations. Apart from the awareness, or rather combined with it, *Developing movement-* is the fixation of the selected effort combination in memory and move- *thinking* ment habit. It is a peculiar kind of building up of ideas about movement qualities and their use. It is perhaps not too bold to introduce here the idea of thinking in terms of movement as contrasted with thinking in words. Movement-thinking could be considered as a gathering of impressions of happenings in one's own mind, for which nomenclature is lacking. This thinking does not, as thinking in words does, serve orientation in the external world, but rather it perfects man's orientation in his inner world in which impulses continually surge and seek an outlet in doing, acting and dancing.

Man's desire to orientate himself in the maze of his drives results in definite effort rhythms, as practised in dancing and in mime. Tribal and national dances are created through the repetition of such effort con- *Effort rhythms and con-* figurations as are characteristic of the community. These dances show the *figurations of social* effort range cultivated by social groups living in a definite milieu. The *their characteristic traits* languid, dreamlike dance of an Oriental, the proud, fierce dance of a Spaniard, the temperamental dance of the southern Italian, the well-measured round dance of the Anglo-Saxon are examples of the effort manisfestations selected and fostered during long periods of history until they have become expressive of the mentality of particular social groups. An observer of tribal and national dances can gain information about the states of mind or traits of character cherished and desired within the particular community. Formerly such dances were one of the main means of schooling the young to adapt themselves to the habits and customs of their forbears. They are in this way as much connected with education as *Dancing developed* with ancestor worship and religion. *movement traditions in*

The urge to play and dance has thus developed into an astonishing *various human activities* variety of movement traditions in all fields of human activity. Dance has *of an ordering principle* been used as a pleasurable aid to work, especially in rhythmical teamwork. Dance has become an adjunct to fighting, hunting, loving and much else. In dancing, or movement-thinking, man first became aware of a certain order in his higher aspirations towards spiritual life. He unconsciously learnt of both the contradicting and the balancing factors within his actions, but he did not know how to use and to control them.

In religious dances man represented those superhuman powers which, as he conceived, directed the happenings of Nature and determined his personal and the tribal fate. He gave physical expression to certain qualities he noted in the actions of these superhuman powers. In such personifications of effort actions, primitive man learnt the reconciling trend of events, and in his movement-thinking he pictured the power behind it all as a god with gliding gestures. Gliding* is essentially a sustained and direct movement with gentle touch. In gliding man and his deity are enveloped in the experience of the infinity of time and the cessation of the drag of weight, but they are actively concerned with the directional clarity of their movements. Many dances of the aborigines of Africa, Asia, Polynesia and America show this feature of gliding in their dance rituals; and the pictures and statues of their gods are represented as figures making gliding gestures. Gods floating over the waters show in ritual, or pictorial representation, a yielding attitude towards the motion factors of Time, Weight and Space. Floating* is sustained gentle and flexible movement, mirroring a state of mind of a similar content.

The malign gods of death and violence are figures represented with thrusting*, piercing and compressing effort actions, all of which are firm and direct movements sometimes occurring suddenly and at other times gradually.

The glittering divinities of joy and surprise are often characterised in dances by flicking* and fluttering movements. Here sensation of lightness is wedded to an indulging in space, which is shown in the flexibility and plasticity of the movements. Sudden appearances and disappearances give the flicking-fluttering movements their brilliance.

Gods of primitive man as symbols of various effort actions

Gods as conceived by primitive man were the initiators and instigators of effort in all its configurations. They were more: they were symbols of the various effort actions. There were dances representing wringing* gods whose movements would be flexible and speak of the gradual passage of time, yet they would be strong and firm. There were slashing* gods who would fight against time and weight with swiftness and powerful resistance and yet be flexible in space, that is readily adaptable to change of shape.

Sprites and goblins, whose movements are imagined to be sudden and direct and yet gentle, are often characterised in dabbing* dances.

The strange poetry of movement that has found expression in sacred dance enabled man to build up an order of his effort actions, which, in essence, is valuable and understandable to this day. Man alone has become

* The significance of these terms is explained later.

aware of the gods. That is to say, man is the only living being who is aware of and responsible for his actions; and he has thus become king of creatures and lord of the earth. From effort awareness in ritual and national dances arose the conventions of the various forms of economic and political order in human society. In the teaching of children and the initiation of adolescents, primitive man endeavoured to convey moral and ethical standards through the development of effort thinking in dancing. The introduction to humane effort was in these ancient times the basis of all civilisation.

Effort awareness and effort thinking in dancing formed a basis for civilisation to progress

But for a very long time man has been unable to find the connection between his movement-thinking and his word-thinking. Verbal descriptions of movement-thinking found their expression only in poetical symbolism. Poetry, descriptive of the deeds of gods and ancestors, was substituted for the simple expression of effort in dance. The scientific age of industrial man has yet to find ways and means to enable us to penetrate into the mental side of effort and action so that the common threads of the two kinds of thinking can finally be re-integrated in a new form. The old ways of effort awareness and effort training will certainly play a part in the investigation of the actor's movement. It is to be expected that mime as expressive of effort, and a fundamental creative activity of man, will, after its long period of neglect, become once more an important factor of civilised progress, when its real sense and meaning have been re-acquired. The value of characterisation through dancelike mime movements lies in the avoidance of the simple imitation of external movement peculiarities. Such imitation does not penetrate to the hidden recesses of man's inner effort. We need an authentic symbol of the inner vision to effect contact with the audience, and this contact can be achieved only if we have learned to think in terms of movement. The central problem of the theatre is to learn how to use this thinking for the purposes of the mastery of movement.

Ways to integrate word- and movement-thinking have yet to be found

Dance-like mime could be an important factor in the attainment of the purposes which the mastery of movement serves

CHAPTER 2

Movement and the Body (Part I)

Flow of movement connected with central and peripheral initiation in the body

The flow of movement is strongly influenced by the order in which the parts of the body are set in motion. We can distinguish an unhampered or "free flow" and a hampered or "bound flow". Movements originating in the trunk, the centre of the body, and then flowing gradually out towards the extremities of the arms and legs are in general more freely flowing than those in which the centre of the body remains motionless when the limbs begin to move. Certain elementary actions have a natural tendency towards "free flow", for example, *slashing* in which the flow of movement is suddenly and energetically released; others, for example, *pressing*, require restraint of flow so that the movement can be stopped at any given moment. It should be noted particularly that a slashing movement of the upper extremities originates centrally in the trunk proceeding towards the shoulders, the upper parts of the arms, and ending in the forearms and hands, while the bound flow of pressing starts in the hands, the tension connected with the action of pressing spreading inwards, first to the wrists and forearms, and then to the upper part of the arms, the shoulders, and finally to the centre of the body and trunk. Even in a light flick of the hand in which the centre of the body seems to remain passive, the energy streams outwards from the forearms to the wrists and hands and finally to the fingers; while a light pressing-gliding of the fingers over a surface will first be felt in the finger tips, then stream inwards into the hands, and so through the wrists to the forearms.

The control of the flow of movement is therefore intimately connected with the control of the movements of the parts of the body. Body movements can be roughly divided into steps, gestures of arms and hands, and facial expressions. Steps comprehend leaps, turns, and runs. Gestures of the extremities of the upper part of the body comprise scooping, gathering, and strewing, scattering movements. Facial expressions are connected with movements of the head which serve to direct the eyes, ears, mouth and nostrils towards objects from which sense impressions are expected. Spine, arms and legs are articulated, i.e. subdivided by joints. The articulation of the spine is more complex than that of the arms and legs.

It is relatively easy to observe a movement flowing freely in an outward direction from the centre of the body through the articulations. The flow of movement is bound when the feel of it takes an inward direction, starting at the outer ends of the extremities, and progressing towards the centre of the body. But there exists a maze of combinations which cannot be demonstrated in a few words. It becomes necessary, therefore, to make a systematic survey of the main types of bodily actions.

The astonishing structure of the body and the amazing actions it can perform are some of the greatest miracles of existence. Each phase of movement, every small transference of weight, every single gesture of any part of the body reveals some feature of our inner life. Each movement originates from an inner excitement of the nerves, caused either by an immediate sense impression, or by a complicated chain of formerly experienced sense impressions stored in the memory. This excitement results in the voluntary or involuntary inner effort or impulse to move.

Bodily movements reveal features of our inner life

Rationalistic explanations of the movements of the human body insist on the fact that it is subject to the laws of inanimate motion. The *weight* of the body follows the law of gravitation. The skeleton of the body can be compared to a system of levers by which distances and directions in *space* are reached. These levers are set in motion by nerves and muscles which furnish the strength needed to overcome the weight of the parts of the body that are moved. The *flow* of motion is controlled by nerve-centres reacting to external and internal stimuli. Movements take a degree of *time*, which can be exactly measured. The driving force of movement is the *energy* developed by a process of combustion within the organs of the body. The fuel consumed in the process is food. There is no doubt about the purely physical character of the production of energy and its transformation into movement.

The switching on and off of the energy and the regulation of the flow of movement according to the intensity of the life-sustaining instinct might also be purely mechanical, but here occurs a break in our rationalistic explanation.

Mechanical explanation of movement not enough

The motion of a falling stone is arrested when it reaches the ground or other support. The acceleration of the speed of its fall and path in space are constant. Both can be exactly calculated. The movement of a dropping arm, however, can be arrested at any time by the control mechanism of the bodily engine. The stop is brought about by purely mechanical means, namely, by the calling into play of an antagonistic muscle which suspends the arm in the air. Yet the cause of the stop is less easy to explain. The falling arm might be stopped because the moving person has realised that there is a dangerous object in the way, and instinct or reflection demands the avoidance of injury and, therefore, the cessation of movement. The individual has accumulated certain experiences of situations and objects which are bound to cause pain or injury and, therefore, he tries to avoid them. It is difficult to ascribe the remembrance of such experiences and the prompt reaction to them to a purely physical or even psychological mechanism. Mechanical theories have been advanced to explain these peculiarities of the behaviour of living beings, but they are not convincing. In any case, it is certain that if a stone would fall straight into a fire, a human being moving in the same direction would try to avoid the fire, by the more or less conscious regulation of his bodily actions.

Movement serves definite purposes

It is a mechanical fact that the *weight* of the body, or any of its parts, can be lifted and carried into a certain direction of *space*, and that this process takes a certain amount of *time*, depending on the ratio of speed. The same mechanical conditions can also be observed in any counter-pull which regulates the *flow* of movement. The use of movement for a definite purpose, either as a means for external work or for the mirroring of certain states and attitudes of mind, derives from a power of a hitherto unexplained nature. One cannot say that this power is unknown, because we are able to observe it in various degrees of perfection wherever life exists.

Man can choose his attitude towards the motion factors

What we can clearly see is that this power enables us to choose between a resisting, constricting, withholding, fighting attitude, or one of yielding, enduring, accepting, indulging in relation to the *"motion factors"* of Weight, Space and Time to which, being natural accidents, inanimate objects are subjected. This freedom of choice is not always consciously or voluntarily exercised; it is often applied automatically without any contribution of conscious willing. But we can observe consciously the function of choosing movements appropriate to situations; that means that we can become conscious of our choice, and can investigate why we so choose. We can observe whether people yield to the accidental forces of weight, space and time, as well as to the natural flow of movement in the sense of having a bodily feeling of them, or whether they fight against one or more of these factors by actively resisting them.

The variety of human character is derived from the multitude of possible attitudes towards the motion factors, and certain tendencies herein can become habitual with the individual. It is of the greatest importance for the actor-dancer to recognise that such habitual inner attitudes are the basic indications of what we call character and temperament.

Effort, the inner aspect of human movement, explained

In order to discern the mechanics of motion within living movement in which purposeful control of the physical happening is at work, it is useful to give a name to the inner function originating such movement. The word used here for this purpose is *effort*.* Every human movement is indissolubly linked with an effort, which is, indeed, its origin and inner aspect. Effort and its resulting action may be both unconscious and involuntary, but they are always present in any bodily movement; otherwise they could not be perceived by others, or become effectual in the external surroundings of the moving person. Effort is visible in the action movement of a worker, or a dancer, and it is audible in song or speech. If one hears a laugh or a cry of despair, one can visualise in imagination the movement accompanying the audible effort. The fact that effort and its various shadings can not only be seen and heard, but also imagined, is of great importance for their representation, both visible and audible, by the actor-dancer. He derives a certain inspiration from descriptions of movements that awaken his imagination.

Descriptions of movement awaken actor-dancer's imagination

In historical descriptions of movements which reach back into the fourteenth century stress is laid on the theatrical performance: the exact keeping of time; the exact remembering of details, gestures and steps; the harmonious co-ordination of the movements of arms and body; the rules of deportment; the imaginative variability of certain details; and last but not least, the significance of floor patterns along which the dancer moves. Such are the contents of these descriptions.

The mode or style of movement used by the dancer or actor sheds a light on a particular aspect of bodily expression. This aspect is perhaps more important in the art of dancing than in mime or drama. The dancer's body follows definite directions in space. Directions form shapes or patterns in space. In fact, dance can be considered to be the poetry of bodily actions in space. In dance a few significant bodily actions are selected, which form the characteristic patterns of a particular dance. Dance patterns can, but need not, have a perceptible mime content. Their significance is not always a dramatic one; it is, indeed, frequently musical, being influenced by the structure and emotional content of the accompanying music. Therefore, the visible movements of the body in the so-called musical dance engender in the spectator reaction of feeling. In dance with dramatic content the spectator's participation is aroused in action and reaction and in the outcome of the conflict. The visible patterns of dance can be described in words, but its deeper meaning is inexpressible verbally.

Various contents of dance

Many dance creations have perished because it was impossible to describe them in words and we hardly know how the dances of former epochs were presented. Had descriptions been attempted, as in fact they have, they could only have been superficial and inadequate. Here the discoveries of industrial man might help the dancer. Modern work analysis

* See footnote on page 9.

*Notation of movement
and dance*

and its notation are not very different from that required to describe any expressive movement. Some modern ballets, for example, creations of the Ballets Jooss and of Balanchine, are recorded in such a notation which developed from traditional attempts to communicate dances by written symbols and is based on observation and analysis of movement, in space and time.

A literature of dance and mime in movement symbols is as necessary and desirable as the historical records of poetry in writing, and of music in musical notation.

*Training of the body
as an instrument of
expression*

The following description of bodily actions is intended to give the student of movement an introduction to exercises designed to train the body as an instrument of expression. In this it is important not only to become aware of the various articulations in the body and of their use in creating rhythmical and spatial patterns, but also of the mood and inner attitude produced by bodily action. How to train the imagination at the same time is sometimes indicated in the text. The reader who takes the trouble to work out and experience the described movements in bodily performance will, however, find his imagination stimulated by doing this.

While it would have been desirable to write the exercises solely by kinetographic means, i.e. by means of movement symbols, for obvious reasons verbal descriptions have been predominantly used. The reader may, however, remember the inadequacy of such descriptions, and in order not to deflect from the main aim, namely that of developing precision of observation and analysis of bodily actions, a manner of description has been attempted which is likely to help the thinking in terms of movement.*

---

* "Kinetography" is a method of movement notation developed by the author. In his book *Principles of Movement and Dance Notation*, also published by Macdonald & Evans, second edition 1975, he explains the system which in its aim has a certain similarity to the universally accepted form of music notation in that whole movement sequences and dances can be recorded and read back. Ballets, folk dances, and any stage movement, as also industrial operations written in this manner, can be completely reconstructed by anyone who might never have seen the original movements, if he knows the principles of Kinetography.

The most comprehensive survey of Kinetography has been worked out by the author's former pupil and colleague, the late Albrecht Knust, who was, until his death in 1978, the outstanding scholar and practitioner in this field. He published the *Handbook of Kinetography Laban*, now expanded into the *Dictionary of Kinetography Laban (Labanotation)*, Macdonald & Evans 1979, which gives concise information about the many possibilities of bodily actions and their notation. Ann Hutchinson, Honorary President of the Dance Notation Bureau, New York, has also published text-books under the title of *Labanotation* in which the original symbols of Kinetography are introduced and their use explained.

ANALYSIS OF SIMPLE BODILY ACTIONS

Before attempting an analysis of bodily actions it may be helpful to consider several sequences which contain typical movement ideas. In relation to these the actions of the body, an extremely versatile instrument of expression, may become more comprehensible.

Each of the following six examples contains a characteristic action mood which may be lyric, solemn, dramatic, jocular, grotesque, serious, or suchlike. How to interpret the sequences is left to the reader's own imagination. It need be mentioned only that the created action-moods will become manifest:

(a) through the particular way the instrument, the body, is used;
(b) through the directions the movements take and what shapes they create;
(c) through the rhythmical development of the whole sequence and the tempo in which it is executed;
(d) through the placement of accents and the organisation of phrases.

Six Examples of Movement Sequences

1. Running — tossing — crouching — whirling — standing.
2. Bowing — lifting — closing — opening.
3. Swaying — circling — spreading — hovering.
4. Trembling — shrinking — precipitating — sprawling.
5. Waving — drooping — perching — pouncing — creeping.
6. Walking — reclining — turning — jumping — uprearing.

These sequences of movement actions create movement events of differing dynamic character. It is important to clarify the transitions between the single actions as only through them the whole tale of the movement event is told. Each sequence can be expressed by an endless variety of action moods arising from the particular mixture of effort qualities with which the actions and the transitions between them are performed. The movement event is realised through bodily actions.

Bodily actions produce alterations of the position of the body, or of parts of it, in the space surrounding the body. Each of these alterations takes a certain time, and requires a certain amount of muscular energy.

It is possible to determine and to describe any bodily action by answering four questions.

(a) Which part of the body moves?
(b) In which direction or directions of space is the movement exerted?
(c) At what speed does the movement progress?
(d) What degree of muscular energy is spent on the movement?

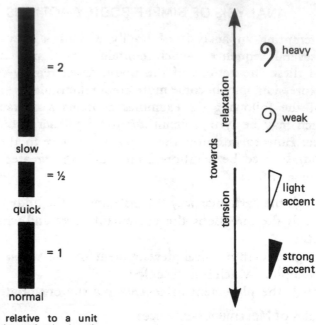

Speed is relative to a unit
which is shown in the length
of the sign.

Let us suppose that the answers to these four questions are as follows.

(a) The moving part of the body is — *the right leg*.
(b) The region of space to which the movement is directed is forward. The movement is — *straight*.
(c) The muscular energy spent on the movement is relatively great. The movement is — *strong*.
(d) The speed of the movement is rapid. The relative pace is — *quick*.

It will be understood that the movement described above is not a step but a thrusting kick of the right leg in a forward direction.

It is obvious that certain subdivisions of the body (*see* Table I), together with certain aspects of the motion factors Time, Weight and Space must be used in order to make observation as exact as possible. Movement is, however, more than the sum of these factors. It has to be experienced and comprehended as an entirety. The urgent advice is given: *invent short movement sequences, or mime scenes, in which the movements described can be recognised.* This is a means of training not only observation but also movement imagination, and of finding the immediate connection with the practical application of bodily exertion in terms of artistic expression.

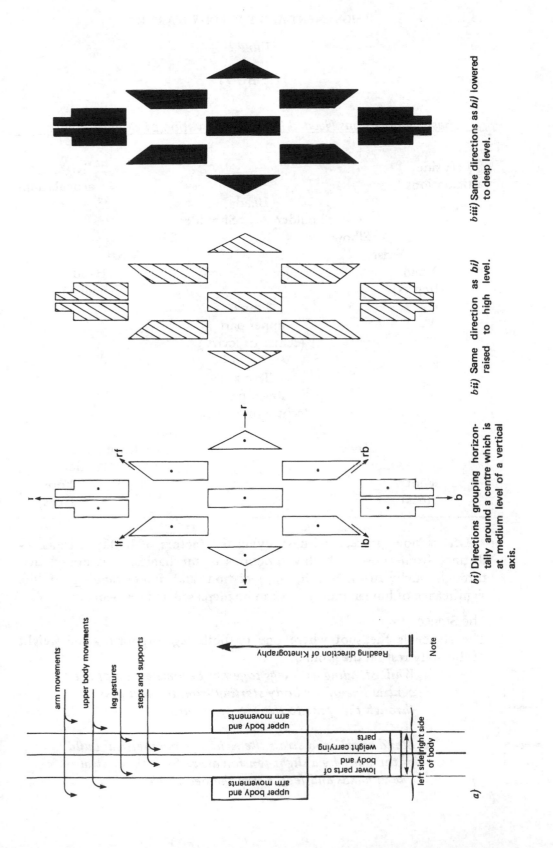

biii) Same directions as *bil* lowered to deep level.

bii) Same direction as *bil* raised to high level.

bi) Directions grouping horizontally around a centre which is at medium level of a vertical axis.

a)

Reading direction of Kinetography

Note

arm movements

upper body movements

leg gestures

steps and supports

upper body and arm movements

weight carrying parts

right side

lower parts of body and

left side of body

upper body and arm movements

Table I

THE BODY

Basic Subdivisions Needed for the Observation of Bodily Actions

Left side
articulations

Right side
articulations

Head

Shoulder   Shoulder

Elbow   Elbow

Wrist   Wrist

Hand   Hand
(fingers)   (fingers)

Trunk
upper part
(centre of levity)

Trunk
lower part
(centre of gravity)

Hip   Hip

Knee   Knee

Ankle   Ankle

Foot   Foot
(toes)   (toes)

Let us now proceed with studying the factors of bodily actions, as distinct from those of bodily functions or mechanics. It is hoped that through such an approach, appreciation and understanding of the significance of human movement can be increased and deepened.

The Stance

The stance is the spot where one or both legs supporting the weight of the body rest on the ground.

1. *While standing with legs together become aware of the central line of the body starting from the feet passing through the spine to the top of the head.*

2. *Stand astride and stress the right-left symmetrical build of the body by a slight tension away from the central line in hands and feet. Carry the head high.*

Transference of Weight or Steps
Each step creates a new stance.

> 3. *Stride out on your right leg, stride out on your left leg and then on your right and left leg again.*
>
> Do this energetically as if "conquering" the new stance each time. Take care that the body weight is fully transferred with every step, leaving the other leg freely hanging in the air.

Directions and Levels of Steps
The spatial direction of a step is relative to the immediately preceding stance.

> 4. *Starting with the weight on both legs, perform the following steps:*
> *forward on to the left foot*
> *forward on to the right foot*
> *now hold the weight on the right one (i.e. the movement action of the right leg is pausing) and step with the left to it, so that the weight is finally on both legs.*
>
> You have completed a "step pattern" which can be repeated:
> (a) with the same or other side;
> (b) into other directions, such as, backwards.

> 5. *Stand on the right leg, with the left leg held backward off the ground. Then perform the following steps:*
> *left foot forward*
> *right foot next to it*
> *(take care that the left leg carries on its action until it is finally off the ground close by the other leg)*
> *left foot forward.*
>
> Another step pattern is completed, which can be directly repeated without any transitional moves:
> (a) with the other side;
> (b) into other directions, such as, sideways.

> *Stand on right leg, hold left leg to your left side off the ground, perform the following step:*
> *with left leg to left side*
> *with right leg next to it*
> *(observe complete transference of weight)*
> *with left leg to left side.*
> *Repeat this step pattern to the right without transitional moves.*

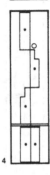

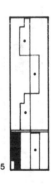

Normal standing and steps are medium level, deep steps are performed by bending knees as much as possible, while the whole of the foot remains on the ground. High steps are those on the ball of the foot with normally extended knee.

Explore the characteristics of deep and high steps which bring about the sensation of either dipping towards the ground or of mounting above it.

In the following exercise you will experience a gradual rising movement followed by an abrupt transition downwards from which the rising phase begins again, each time into a new direction.

6. *Stand on both legs, step into the directions*
   right-forward with the right leg deep
   right-forward with the left leg medium
   right-forward with the right leg high

   left-backwards with the left leg deep
   left-backwards with the right leg medium
   left-backwards with the left leg high

   right-backwards with the right leg deep
   right-backwards with the left leg medium
   right-backwards with the right leg high

   left-forward with the left leg deep
   left-forward with the right leg medium
   left-forward with the left leg high.

   Complete by lowering the weight on the left leg to medium level, at the same time stepping with right leg next to it, so that the starting position is again attained.

The whole exercise can be repeated:

   (a) exactly the same;
   (b) in the same way but starting into any of the other three diagonal directions;
       left-forward,
       left-backward,
       right-backward,
   (c) reversing the level of steps, beginning high-medium-deep;
   (d) varying the levels otherwise.

When stepping diagonally observe that the front of the body is not changed, i.e. the hips should retain their

6

original position so that open and crossed positions of the
legs occur, giving experience of either a free or a hampered
way of striding.

Gestures with Successive Actions

Gestures are actions of the extremities which do not involve transference
or support of weight. They can occur towards, away, or around the body,
and can be done with *successive* actions of the various parts of a limb.

7. *Stretch right arm away from your body. Fold it in
   towards your chest.*

   As you are doing this become aware of the different
   articulations of your arm; when stretching arm out for
   starting position make sure that your hand and all the
   fingers are stretched too. Now move first only the hand
   towards your body, then the wrist, followed by the elbow
   and finally by the shoulder. In this way you will perform
   a *successive* movement of your right arm, starting from the
   fingertips which move the hand, followed by the wrist
   which moves the forearm, then the elbow moves the upper
   arm, and finally the shoulder moves the right shoulder
   girdle, thus completing a folding action towards the central
   line of your body.

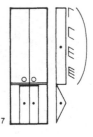

7

   (a) *Reverse the action to unfolding, moving away
       from the central line into space successively:*
       shoulder — *affecting shoulder girdle*
       elbow   — *affecting upper arm*
       wrist   — *affecting forearm; and finally*
       fingers   — *affecting hand.*
   (b) *Repeat folding and unfolding with the left arm.*
   (c) *Fold and unfold both arms, observing a clear
       succession in the action of the various articula-
       tions of your arms by altering their position.*
   (d) *Fold and unfold one or both arms with different
       arrangements in the succession of action of the
       various arm articulations, e.g. stretch arm away
       from body. Then fold it towards the body with
       an irregular succession of first shoulder, then
       hand, elbow, and finally wrist.*

   Notice the more harmonious effect of folding and unfolding when the action is taken in regular succession of
   the arm as shown above — no matter whether starting with
   the inner or outer articulations first, while the movement
   is more disharmonious, even grotesque, when the succession
   is irregular, as in Exercise 7 *(d)*.

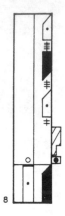

8. *Fold and unfold a leg. When weight is carried on one leg, the other is free to perform gestures. Stretch one leg away from the body in the air. Perform a successive folding movement as follows:*

    hip   — tilting the pelvis
    knee   — moving the upper leg
    ankle — moving the lower leg
    toes   — moving the whole foot

thus accomplishing a folding of the leg towards the central line of the body.

  (a) Reverse the succession to unfolding, beginning with the toes and ending with the hip.
  (b) Fold starting from tip of toes.
  (c) Unfold starting from hip.
  (d) Perform movements of folding and unfolding with irregular succession of parts of the leg.

Gestures with Simultaneous Actions

In contrast to the previous exercises, movements of the extremities towards the body and away from it can also be performed with *simultaneous* actions of their various articulations. Observe that thereby the movements of hand, forearm, upper arm, shoulder-girdle, or those of foot, foreleg, upper leg, pelvis, are started and completed at the same time.

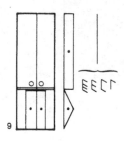

9. *Repeat Exercise 7, but instead of completing the movement of each part of the arm before the next one starts, all four are in action simultaneously from beginning to end.*

    Compare the experience of simultaneous and successive movements. You will notice that the flow of the successive action enhances awareness of the process of folding and of unfolding, while in simultaneous action it is the achievement at each stage which seems to have importance.

10. *Experiment with successive and simultaneous folding and unfolding of arms and legs using only two or three of the articulations, e.g. shoulder and wrist, or knee and ankle.*

  (a) Move both arms and one leg at the same time, varying simultaneous and successive actions within each limb.
  (b) Fold or unfold one limb after the other, varying simultaneous and successive actions.

Directions and Levels of Gestures

Directions and levels of arm and leg gestures are relative to the joint in which movement takes place.

Medium level of a leg gesture is at hip height, medium level of an arm gesture is at shoulder height. Movements reaching above these joints are high, those reaching below are deep.

Gestures leading into the vertical line below, above or close to the moving joint of the limb are deep, high or medium respectively.

Arms hanging down vertically from the shoulders as in normal carriage are "deep". A leg hanging down next to the other leg which supports the body weight is "deep".

Arms raised vertically above the shoulders are "high", legs raised vertically above the hips, as in a handstand, are "high".

Arms or legs closely folded near shoulders or hips respectively are at "medium" level.

Specific levels and directions of the single parts of arms or legs are related to the joints in which the movement takes place. For instance, a movement of the knee forward-medium brings the thigh forward to hip level, a simultaneous movement of the ankle backward-deep, that is, backward-deep from the knee joint, brings the foot to near the knee of the supporting leg.

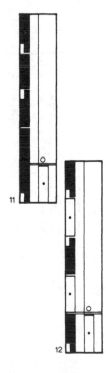

11. *Stand on right leg, extend left leg backward-deep.*
    Action of the whole left leg:
        deep
        forward-deep
        deep
        backward-deep.
    Practise the same action with other leg.

12. *Stand on right leg, extend left leg backward-deep.*
    Action of the whole left leg:
        medium
        forward-deep
        medium
        backward-deep.
    Practise the same action with other leg.

    While the free leg performs gestures, the weight-supporting leg can change level.

13. *Stand on right leg deep, extend left leg behind diagonally across to right-backward-deep, touching floor with toes, but not carrying any weight. Perform the following leg actions at the same time.*

Left leg gesture:	*Right leg on stance:*
via medium	*high*
to left-forward-deep	*deep*
(with touch of floor)	
via medium	*high*
to right-backward-deep	*deep*
(with touch of floor)	
via medium	*high*
to left-forward-medium	*deep*

(a) *Repeat the whole sequence with other side.*

(b) *Repeat the sequence starting left leg gesture: left-backward-deep (i.e. in an open diagonal backward position).*

(c) *Reverse the sequence by starting with left leg gesture: right-forward-deep, or left-forward-deep.*

Leg Gestures May Precede a Step

Gestures may be done as a preparation to a transference of weight. Perform the following example.

14. *Starting position on both feet together: medium level.*
Right leg gesture:
    rightside-deep, forward deep.
Right leg step:
    leftside-deep.
Then change level to medium while stepping with the left foot next to the right one, also medium level. Repeat with other side.

Leg Gestures May Follow a Step

These gestures are felt as a result of the step, as, for instance, after a left step backward-deep the right leg performs a folding movement across the left knee.

Leg gestures may also occur on their own, any transference of weight which may eventually follow being incidental. See Examples 11 and 12, or the following.

15. *Stand on left leg deep.*

   (i) ⎰ *Right leg gesture:*
        knee rightside-medium
        ankle leftside-deep (bringing right heel near
           left knee)
        toes rightside-medium.

   (ii) ⎰ *Change of level of supporting leg to medium*
        ⎱ *right leg gesture to rightside-medium.*

   (The braces indicate simultaneous actions.)

   (a) *Repeat several times.*
   (b) *Repeat with other leg.*

Gestures via Several Directions Constitute Definite Shapes of Movement

Among the spatial forms of movement we can distinguish round, angular and twisted shapes as being basic.

16. *Left arm gesture:*
    from right-forward-medium
    to   forward-deep
    to   left-forward-medium
    to   forward-high, return
    to   right-forward-medium.

   This sequence which involves only gradual changes of direction creates an almost *round* spatial pattern of shape, using the forward direction as its centre.

   Gestures of arms and legs should be executed with ease and not with any stiffness in their various articulations.

17. *Right arm:*
    rightside-deep
    left-forward-deep
    rightside-medium
    left-forward-high
    rightside-high.

   This sequence of directions, abruptly changing, produces a zig-zag kind of pattern with *angular* shapes.

18. *Perform both arm gestures at the same time.*

Left arm gesture:	*Right arm gesture:*
rightside-medium	*leftside-medium*
forward-medium	*forward-medium*
deep	*deep*
leftside-medium	*rightside-medium*

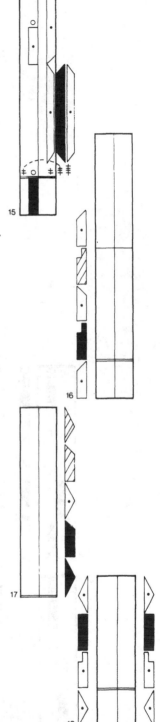

A *twisted* shape occurs when the line of movement, with
gradually changing directions, embraces two centres. In
this example the left arm winds around the centres:

right-forward-deep and
left-forward-deep;

while the right arm winds around:

left-forward-deep and
right-forward-deep.

Round, angular and twisted shapes call for clear bodily execution.
Observe how the functions of the body, namely those of bending, stretching, twisting, and of their combinations operate in performing these
shapes.

Combined Arm and Leg Movements

The body is our instrument of expression through movement. The body
acts like an orchestra in which each section is related to any other and is
part of the whole. Its various parts can combine in concerted action, or
one part may perform alone as a "soloist" while others pause. It is also
possible that one or several parts take the lead and others accompany.
This is important to remember when studying the exercises in this
chapter. Each action of a particular part of the body has to be understood
in relation to the whole which should always be affected, either by participating harmoniously or by deliberately counteracting or by pausing.

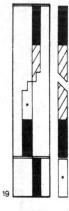

19. *Starting position: weight on right leg deep, right arm*
    medium. Perform arm and leg movements at the same
    time.

legs:	Right arm:
left step: forward	*gesture: left-forward*
(via deep to medium)	*(via deep to high)*
right step: backward	*gesture: right-backward*
(via high to deep)	*(via high to deep)*

Change of Level

Change of level can occur during one single action of the body or its
parts. In the preceding exercise the transference of weight in the stepping
movements is completed only when the knee — after bending — is again
stretched. This produces a smooth stepping movement, different from the
following:

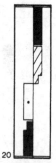

20. *Left step: forward-deep*
    medium.
    Right step: backward-high
    deep.

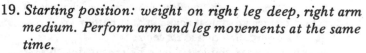

Practise complete transference of weight with the deep step forward and, after a hardly perceptible pause, stretch supporting knee.

Similarly, practise complete transference of weight on to right foot backward-high and, after a hardly perceptible pause, lower the heel and bend knee.

The arm gestures in Exercise 19, which change level while moving left-forward and right-backward, also produce smooth curves different from the following exercise in which each direction is reached by an action of its own.

21. *Right arm gesture:*
    left-forward-deep
    left-forward-high
    right-backward-high
    right-backward-deep.

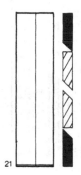

Train your appreciation of the difference between change of levels within one action and change of directions brought about by different body actions.

Extension of Gestures

The normal reach of our limbs, when they stretch away from our body without changing stance, determines the natural boundaries of the personal space or "kinesphere" in which we move. This kinesphere remains constant in relation to the body even when we move away from the original stance; it travels with the body in the general space.*

Gestures can reach far away from the body or near to it, or travel between near and far. Sometimes they are restrained and close, at other times extended and wide involving movement of the whole torso, thus causing the kinesphere to shrink and to grow.

Gestures linking different extensions of the kinesphere are often either exceedingly controlled and static or tend to be charged with dynamic qualities often leading to locomotion.

22. *Right arm from medium to narrow forward-deep (narrow or near the body means the arm will not be fully extended), to wide forward-upward.*
    Perform this curved gesture:
    (a) with a freely and swiftly flowing movement so that its extension beyond easy reach causes the whole body to overbalance and necessitates a step or two in order to re-establish balance;
    (b) with a determination to uphold the original stance. This will result in an extreme extension of

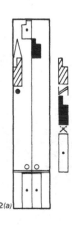

* See *Modern Educational Dance* by Rudolf Laban, Macdonald & Evans, third edition 1975.

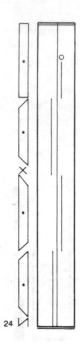

the arm and in a feeling of boundness and control.

23. *Explore the possibility of arm gestures leading to other directions and levels in the extended kinesphere, such as:*
    (a) high, causing the feet either to leave the floor freely, or to clutch the floor in an attempt to counteract the pull into an extended sphere;
    (b) right-backward-deep, leading to falling or to controlling the balance.

24. *Left arm gesture:*
    from   *leftside-medium (wide)*
    to   *rightside-medium (normal)*
    to   *leftside-medium (narrow)*
    to   *medium.*

    This sequence of movements creates a swaying motion gradually leading to a standstill. Adapt the position to your feet in such a way that the decreasing extension of the arm gesture can be clearly felt.

25. *Invent arm gestures creating curved or angular shapes of movement while travelling in one or between the different extensions of the kinesphere, such as:*
    Right arm gesture:
    from   *forward-medium*
    to   *rightside-medium (wide)*
    to   *deep (narrow)*
    to   *backward-medium (wide).*
    Or
    Left arm gesture:
    from   *rightside-high (narrow)*
    via   *forward-high (narrow) and*
           high (narrow)
    to   *forward-deep (wide)*
    to   *right-backward-medium.*

    Help to bring out the changing extension of your arm gesture with appropriate movements of your legs and torso. Observe how the change of extension in space calls for an increased sensitivity in bodily actions.

26. *Experiment with leg gestures in a similar way as outlined for the arms.*

Extension of Steps

As indicated before, leg gestures may be done by themselves or in connection with a step, either preceding or following it. Steps too can be narrow and wide, or better small and big, yet they do not increase or decrease the kinesphere but cause the personal space to enter the general one.

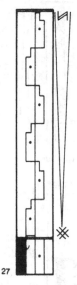

27. *Execute a number of steps leading forward, starting with very small steps gradually increasing to very wide extended ones.*

28. *Start with very big steps forward and end with very small ones.*

    When doing Exercises 27 and 28 observe whether any changes occur within each one as to tempo and your use of force. If so try to:

    (a) refrain from any changes and work for an even control of the bodily actions;

    (b) reverse or vary the changes you have noticed without disturbing the gradual increase or decrease of spatial extension.

27

29. *Stand on right leg deep.*

    Left leg gesture:
    from   deep (narrow), that is left foot near right knee
    to   right-backward-deep (wide)
    to   left-backward-deep (wide)
    followed by a very small step on the left foot to
    *   backward-medium.*

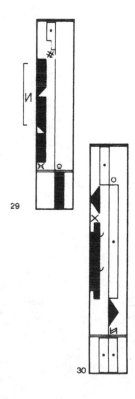

30. *Perform the following leg actions at the same time:*

Left leg:	Right leg:
—	step rightside-deep (very wide)

Gesture:

backward-deep (with touch of floor)	
forward-deep (with touch of floor)	medium
rightside-deep (narrow)	
Step: to stance medium	pause

29

30

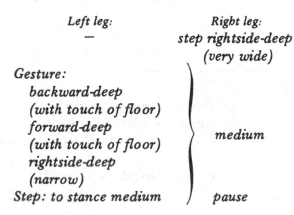

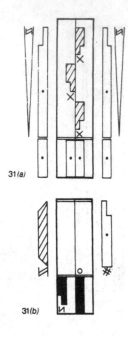

31 *(a)*

31 *(b)*

31. *Invent sequences of leg movements with narrowing and widening gestures preceding or following small or big steps; accompany with arm gestures, e.g:*

    (a) *three small steps backward-high while both arms starting from medium gradually extend to forward-medium very wide;*

    (b) *stand on right foot deep while the left leg is extended far backward-deep, touching the floor with the big toe. Now perform both arm gestures at the same time.*

Left arm gesture:	*Right arm gesture:*
to leftside-high	*to right-forward-medium*
(very wide)	*(very narrow)*

Observe how simultaneous actions of different parts of the body, which vary in their extension, create expressive content of what one might call a problematic nature, i.e. they do not suggest a straightforward solution.

Table II

SPACE

Elementary Aspects Needed for the Observation of Bodily Actions

Directions:	
	forward
	left-forward right-forward
	left right
	left-backward right-backward
	backward
Levels:	
	high
	medium
	deep
Extensions:	
	near — normal — far
	small — normal — big
Path:	
	straight — angular — curved

Speed

The rate at which we let movements follow one another is the *speed* with which we act. Our ordinary walking pace might be considered to be of medium speed. We can give each step a time unit which may correspond to each beat of our pulse.

A step taking several pulse beats is felt as being slow — and several steps taking only one pulse beat are quick. (*See* Table III, p. 00.)

> 32. *Perform a number of steps in any direction you like at approximate rate of your pulse, i.e. medium or "normal" speed.*

> 33. *While shifting your weight from one foot on to the other let several pulse beats pass by and perform thus a "slow" step.*

> 34. *Try to make several steps between two pulse beats by acting "quickly".*

Time-rhythm

The time-rhythm of a series of movements consists of a combination of equal or different lengths of time units. These can be represented by the musical signs of time values.

> 35. *Create step-patterns using the following rhythms repetitively:*
>
> (1) ♩. ♪ ♩. (2) ♩ ♩ ♩ (3) ♩ ♩ ♩ (4) ♪ ♪ ♪
>
> (a) *Compare the step-patterns you have created with one another and stress the characteristic bodily actions of each.*
> (b) *Produce variations of each step-pattern by introducing leg gestures without changing the original rhythm.*
> (c) *Invent sequences of arm gestures to each rhythm with clear use of the various articulations (shoulder, elbow, wrist, fingers) in either a simultaneous or successive manner.*
> (d) *Observe changes of directions and levels in relation to each rhythm and produce variations through replacing them by their opposite, i.e. forward is replaced by backward, upward by downward, etc.*

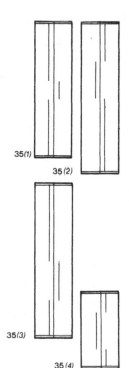

35(1) 35(2) 35(3) 35(4)

Such exercises will not only increase precision but also
ability for invention of bodily action.

Tempo

Time-rhythms are independent of the tempo of the whole movement
sequence. The same rhythm can be performed in different tempi without
changing the proportional length of each time unit.

36. *Perform your movement sequences created in con-*
    nection with the rhythms given in Exercise 35 at
    different tempi:
    (a) medium, (b) slow, (c) quick.
    Ex. 35 (1) expressed in musical signs would read:

Pause

Any bodily action can be arrested and held for a period of time. The
length of the pause can be measured by time units which are proportion-
ate to those of the movements introducing and concluding the period of
stop.

37. *Choose any initial position; change this by a bodily*
    action taking three pulse beats (or three even counts);
    arrest this action by pausing for two time units in
    whatever position you have arrived at and become
    aware of the nature of this new position. Then return
    to the original taking one time unit and without pausing
    start the whole phrase again.

    (a) Start with pausing in an initial position for three
        time units; change into another position (precon-
        ceived) over two time units; change this back into
        the initial position in one time unit.

    (b) Change an initial position over three time units.
        Change the newly gained position into an entirely
        new one during two time units and hold this one
        for one time unit. Repeat this rhythm, each time
        producing actions in different parts of the body
        and into different directions and levels.

38. *Invent sequences of bodily actions using variations of quick, slow and medium speeds with and without:*
    (a) arresting an action;
    (b) pausing after a completed action in a new position over a period of time clearly proportionate to that of the action.

Vibratory Shaking Movements

These are very quick changes of a number of positions produced within one time unit.

39. *Produce a vibratory shaking movement:*
    (a) by very quick changes of two alternate positions;
    (b) by very quick changes of many different positions.

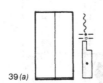

39 (a)

It will be apparent that the position of such shaking movements will be situated close to one another in space, e.g. your right arm stretched out forward moves quickly a number of times a small distance up and down during one time unit, or your knees move very quickly inward and outward over a period of time.

Table III
TIME
Elementary Aspects Needed for the Observation of Bodily Actions

Speed:	quick			normal			slow		
(Time units)	1	1½	2	3	4	6	8	12	16
Tempo: (related to movement sequences)	presto			moderato			lento		

The Use of Muscular Energy or Force

In order to bring about a change in our bodily position we use muscular energy. The deployment of strength and its degrees is in proportion to the weight carried or to the resistance given to it. The weight can be either:

(a) that of the body part moved; or
(b) that of an object being moved.

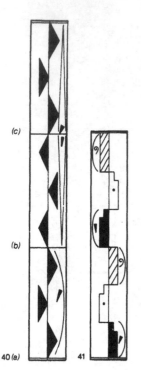

The resistance can be produced either:

    (a) from within one's own body by antagonistic muscles; or
    (b) from without by objects or persons.

Resistance can involve strong, normal or weak muscular tension.

40. *Perform a series of steps into one direction:*
    (a) with equally strong resistance against the floor;
    (b) with increasing strength;
    (c) with decreasing strength;
    and invent variations of time-rhythms.

41. *In a sequence of three steps perform the first one with strong, the second with normal and the third with weak muscular tension. Repeat several times. Create one or several repetitive dance steps by giving to each step particular direction and level.*

42. *Invent single bodily actions, such as occur in the unfolding of an arm, the bending of the knees, etc., and experiment within each one with changes of strong, normal and weak muscular tension.*

43. *Perform a continuous arm gesture from leftside-medium to high to rightside-medium to forward-medium in which the movement into the first direction has strong, into the second normal, and into the third weak tension.*

Accent

A suddenly or gradually arising tension may produce a stress or accent for a rhythmically important movement.

44. *Perform a series of stepping movements giving an accent to:*
    (a) every fourth one;
    (b) every second one.

    It is interesting to observe that the one-sided stress in the body, i.e. the accent occurring each time on the same leg, gives either an awkward or an exhilarating feeling, as the case may be, mainly in example *(b)*. This may be ascribed to the equal alternation of legs as well as of stress and non-stress, while in example *(a)* the unstressed period is brought about by several steps giving the action better balance.

In contrast to this notice that the special stress on each third step in Exercise 41 happens on alternate legs, bringing about a comfortable feeling or maybe one of complacency. This is particularly noticeable when the three steps are performed with even speed, as in a lilting waltz.

45. *Vary the placing of accent within a sequence of three steps in the following ways:*
(a) on the first step;
(b) on the second step;
(c) on the third step;
at first repeating each example four times, later establishing longer phrases mixing the examples.

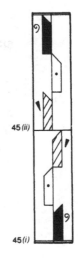

45 (ii)

45 (i)

In the alternation of accented and unaccented movements we can distinguish, for instance, the following two possibilities of phrasing:
> *(i)* the unaccented part precedes the accent and leads up to it;
> *(ii)* the unaccented part follows the accent and, so to speak, dissolves it.

46. *Perform Exercises 44 (a) and (b) with definite awareness, whether you lead up to the accented step, or whether you start with it and round off the phrase with lesser tension. Vary the use of directions and levels.*

Table IV
WEIGHT
Elementary Aspects Needed for the Observation of Bodily Actions

Muscular Energy or Force used in the resistance to weight:	strong 2 : 1	normal 1 : 1	weak ½ : 1
Accents: Degrees of tension:	stressed tense	or to	unstressed relaxed

Turns or Changes of Front
In the examples given hitherto the front of your body has remained unaltered in its direction. Now you will practise turning your front into different directions. In order to gain secure judgement of the degree of turn you will have to free yourself from gazing at particular objects in your surroundings; the means are to be found in your own body. As you

stand erect on your two feet, perhaps best with your eyes shut, try to get a clear image of the relations of your body to the spatial directions of:

> forward being in front of body;
> backward being behind body;
> sideways right being at right of body;
> sideways left being at left of body;

and not where the windows, doors or fireplace of a room are.

Therefore, when you change your front into another direction, e.g. backwards, that is where your back was before, you have performed a half turn. After accomplishment of the turn the old backward direction has become the new forward direction. This will be elucidated in the following exercise.

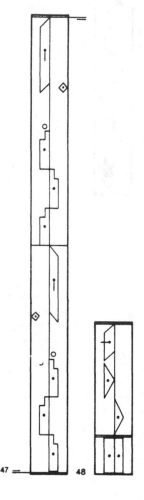

47. *Step forward with the right, with the left, with the right foot, pause and realise that the left leg is held backward — let the big toe touch the floor — now turn to your left without moving the left toe away from the spot until the left leg is in front of the body.*

> You will have performed a half turn and the left leg is now in a position ready to repeat the sequence and to retrace the original path-way.
>
> Notice that at the end of this the right leg is behind with toes touching the floor, and in order to repeat the entire sequence from the very beginning you will have to perform another half turn, this time to your right.

48. *Step to the right side first with right leg, then with left leg (crossing over). Now turn to your right until your front faces the direction which was initially at the left of your body. This means doing a three-quarter turn to the right.*

Direction of Turns

The change of front or turning towards a new direction can be effected in two ways:

    (a) to the right, i.e. right shoulder goes backward;
    (b) to the left, i.e. left shoulder goes backward.

Reasons for turns

Turns are frequently performed:

    (a) either in order to make a deliberate change of direction, as in Exercises 47 and 48;
    (b) or to enjoy the swift pivoting action which is usually initiated by gestures of arms or leg, e.g.:

47 ═   48

49. *Swing your left leg swiftly from leftside via forward to rightside and allow your body to be turned around as far as the impetus of the leg swing will take you. Then perform three steps backward.*

Note that "backward" is relative to the new front after the turn.

This turn is a so-called "inward" turn as the gesture initiating it crosses over the front of the body and gives a closed-in feeling. The opposite is the case when the gesture comes from forward and opens out to backward as in the following exercise.

50. *Perform a circle with your right leg swiftly around the left leg which supports the weight, from front via rightside, backward to leftside, and allow the body to pivot.*

This is an "outward" turn and the feeling of openness can be enhanced by accompanying opening arm gestures. Finish this turn with one or two steps of your own choosing.

Placing of Turns

Turns can also occur between steps and different expressive qualities can be brought about by placing them as follows:

(a) turn preceding a step;
(b) turn while transferring weight;
(c) turn following a step.

51. *Experiment with the above exercises, introducing:*
    (a) changes of levels;
    (b) changes of speed.

Change of Front while Progressing Through Space

Change of front can be effected during several steps.

52. *Take several steps backward and keep changing your front by turning to your left until you face the direction which initially was behind you. You will have performed a half circle.*

Experiment with similar change of front while using another constant direction of stepping.

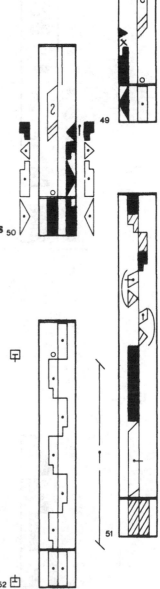

Jumps

One of the most exciting bodily actions is the elevation of the body off the floor, that is, when both your feet leave the floor in a jump and you are actually suspended for a moment in the air.

Jumping requires to a certain extent quick and strong actions of the legs in order to throw the body up in the air.

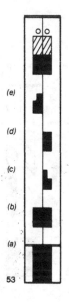

(e)

(d)

(c)

(b)

(a)

53

53. *Experiment with jumping and you will discover that there are five basic possibilities:*
    (a) from both feet on to both feet;
    (b) from both feet on to one foot;
    (c) from one foot on to the same foot;
    (d) from one foot on to the other foot;
    (e) from one foot on to both feet.

    While these are natural variations which each child enjoys when playing hopscotch, in the classical ballet they are executed in a particular style and are termed *(a)* sauté; *(b)* sissonné; *(c)* levé; *(d)* jeté; *(e)* assemblé.

    When leaving the floor with both feet at the same time as in *(a)* and *(b)*, elevation is brought about by the actions of the feet and the knees only.

54. *Explore different ways of jumping off both feet with regard to:*
    (a) levels of standing;
    (b) open or closed position, i.e. feet close together or with a distance between them;
    (c) degree of energy;
    (d) speed of action.

    When pushing off the floor with one leg, as in 53 *(c)*, *(d)*, *(e)*, the other can help elevation by initiating the jump with a vigorous gesture upwards, as outlined in 55.

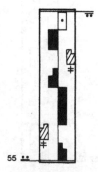

55

55. *Step with right leg forward and immediately throw left knee forward-high so that the right foot is forced to leave the floor for a moment and then let the weight fall back on it. Repeat with other side and continue this "skipping" movement several times.*
    (a) Perform steps into different directions.
    (b) Change your front by turning in the air.

56. *Stand on one leg and hold the other backward-deep*
    in readiness to throw it forward and to jump on to it.
    Continue this "leaping" movement from one leg to
    the other.
    Vary these leaps using different extensions:
    (a) in transference of weight, i.e. near or far distance;
    (b) in leg gestures, i.e. bent or stretched knees.

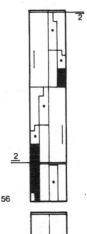

Duration of Elevation

High and wide jumps require more time than low and near jumps.

Running consists of a series of low and not particularly wide jumps.
The speed is relatively quick.

56

57. *Prepare a jump off one foot high up into the air by a*
    short quick run and land on both feet deep. Take for
    each running-step one time unit and for the jumping
    off and the suspension in the air two time units.

58. *Experiment with different kinds of jumps with*
    special reference to quick and slow leg actions:
    (a) with or without turning;
    (b) with or without particular leg gestures;
    (c) on the spot or away from it;
    (d) with preceding or following run;
    (e) with arrested leg actions during flight, i.e. legs are
    *   held in a particular position.*

Combinations of Body, Time, Space and Force Indications

Any bodily action can be understood as using one of the various combinations of the subdivision of body, time, space and muscular energy
previously enumerated.

The very great number of these combinations corresponds to the
possible acts of movement which can be registered in a logical manner.
The order shown in such combinations will best be observed by the movements used in dancing, since they are relatively large and clear, and therefore more easily recognised. The analysis of bodily actions in sport, play,
acting, work, and everyday behaviour is based on the same "thinking in
terms of movement" as is applied for that of dance movements.

It is to be noted that no special method or style of dancing is at the
basis of the logical order of movement observation. Dancers in every age
and in all countries have thought, and still think, in terms of movement
— essentially space, time and energy — indications.

57

For instance, all classical ballet positions, steps and gestures can be described in terms of movement without reference to the conventional names given to them. They are a specific and stylised form of the vast treasure of movements possible for the human body. The same refers to any other form of historic, period or national dance styles, including those of exotic dances. Each style represents a special selection of movements originated from racial, social, period and other characteristics. The free movements of the modern stage artist embrace many possible combinations of bodily actions.

59. *Observe bodily actions:*
   (a) of a person in everyday life;
   (b) of a person portraying a character in a mime scene;
   (c) of a dancer performing a particular national or
       period dance.
   Analyse these from the following points of view.
   (a) Ways of using the body, whether:
       upper or lower part of body;
       right or left side of body;
       off or on the floor;
       symmetric or asymmetric;
       simultaneous or successive movements in one
       or both limbs.
   (b) Space, such as:
       directions and levels of steps and gestures;
       change of front;
       extension of steps and gestures;
       shape of gestures.
   (c) Time, such as:
       quick and slow in gesture and step;
       repetition of a rhythm;
       tempo of a rhythm.
   (d) Weight, such as:
       strong or weak tension;
       placement of accents;
       phrasing arising from stressed and unstressed
       periods.

60. *Other peculiarities of movement can be observed and*
   analysed, namely, that:
   (a) several parts of the body can perform similar or
       different acts of movement simultaneously. So

also can several dancers dancing together perform similar or different acts of movement simultaneously;

(b) *movement can result in touch or grip. Feet, hands, or other parts of the body can touch either objects or persons. Hands can grip.*

The logic which thinking in terms of movement requires should first be developed from the observation and description of simple movements in which neither simultaneity, nor other correlation of limbs, such as touch and grip, arise. The only touch which could be considered from the very beginning is that occurring when first one foot and then the other touches the ground before or after a step.

Table V
FLOW
Elementary Aspects Needed for the Observation of Bodily Actions

Flux:	going	interrupting	arresting
Action:	continuous	jerky	stopped
Control:	normal	intermittent	complete
Body:	motion	series of positions	position

In order to increase our capacity of recognising bodily actions it will be necessary to become acquainted with further variations of the use of the body and its parts together with space, time and weight components. This is done in the following chapter, which will deal in more detail with the fourth motion factor — Flow. This is broken down into its elementary aspects in Table V.

CHAPTER 3

Movement and the Body (Part II)

On entering the second phase of our examination of bodily actions the reader should be reminded that these represent only one of the various considerations which play a part in the art of movement. It is perhaps the most important, since the body is the instrument through which man communicates and expresses himself. Therefore anyone who cultivates this art, but in particular the stage performer, has to acquire ability for distinct bodily action, i.e. clear use of the body and its articulations both in stillness and in motion. The movements of each part of the body are related to those of any other part or parts through temporal, spatial and tensional properties.

In the previous chapter the reader has been introduced to the various elementary aspects of these and he has been led to recognise bodily actions as alterations of the positions of the body or its parts, taking time, occurring in space and using force. An analysis based on motion factors enhances the thinking in terms of movement, while an explanation of the functioning of the body such as bending, stretching, twisting, would tend to stress mechanical rather than expressive awareness.

In this connection reference should be made to the flux of movement which is an aspect of the motion factor of Flow (see Table V). Flux is the normal continuation of movement as that of a flowing stream and can be more or less controlled.

If the flux of bodily actions is completely stopped a position results. If the flux is interrupted intermittently, a trembling kind of movement is produced. Most exercises in the previous chapter called for continuous succession of bodily actions. The reader is advised to recall a few of the movement sequences and to apply consciously a control of their normal flux by:

(a) intermittently interrupting; or
(b) entirely stopping it.

Six Examples of Movement Scenes

The following description of situations, moods and actions may stimulate the reader to create movement scenes in acting, miming, or dancing, as the case may be. After he has done so he might check his bodily actions from the points of view already discussed. He will, however, soon find that no proper guidance has yet been given with regard to movements of the trunk, head and hands. Neither have positions other than standing been considered, namely those of kneeling, sitting, lying, nor situations of the centre of gravity. It is also important to establish a norm by which spatial relations to people and objects can be recognised. Nevertheless, the reader is encouraged to make his own discoveries.

1. Entering the dark room the man perceived a tiny little squeaking sound which made him tingle.

2. The harvesting done, the people gathered in the square to rejoice in singing and dancing.

3. She was tired and sat down to rest. For the first time she was able to survey the situation calmly and when she realised that everyone expected her to take the lead she felt strength and purpose coming into her being, filling her with renewed zest to continue on her way.

4. The vigorous rhythm of the music made them jerk their bodies hither and thither and their feet carried them all over the floor tracing the strangest patterns on the ground.

5. He felt excited in that wide expanse of space and he reached out to touch the ends of the world. But when he did not seem able to hold them he turned to the spot where he was standing, clinging to it, for this was where he would make his dwelling.

6. Only in dreams did she know how to fly and she recalled the bliss she felt when her body was carried up and suspended in the air and then gently lowered to the ground. The lifting and sinking came in endless repetition like the waves of the sea.

ANALYSIS OF COMPLEX BODILY ACTIONS

In the descriptions of the exercises now following reference is made mainly to the movements of the body in space. The reader is, however, reminded that if bodily actions are to reveal (both to the doer and the spectator) features of inner life, the time and weight factors of movement must play their part. It is therefore of greatest importance that the reader pays careful attention to the rhythmic-dynamic patterns which he should develop himself.

The Trunk

The trunk is that part of the body which includes both the pelvis and the shoulder girdle and it is moved from the hip joints. Its actions are defined by the movements of the spine which show great versatility. For the time being, the head should also be considered as part of the trunk.

There are two main aspects which may be useful in the observation of trunk actions.

(a) Carriage. The normal manner of carrying one's body is upright, i.e. the trunk is "high" and above the hip joints. Carriage has mainly to do with the positioning of the trunk which can undergo a great number of variations.

(b) Motion. The trunk can follow, counteract or replace gestures of the limbs, particularly those of the arms.

In either case the trunk actions depend on which parts of the spine participate, whether in a successive or simultaneous manner, how the energy used is distributed, with what speed they are executed, and whether they are done jerkily or fluently.

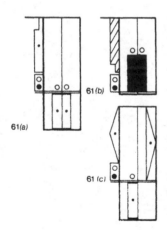

61(a)

61(b)

61(c)

61. *Explore the possibilities of directional movements of the whole trunk (including head) from a standing position. The spine has to be kept immobile.*

> You will find that:
> *(a)* into the area in front this is easily done by bending at an angle in the hip joints; while
> *(b)* into the areas behind the legs will have to participate in the action, at least as far down as the knees, which are bent at an angle;
> *(c)* into the areas immediately at either side one leg will probably be lifted off the ground.

62. *Explore the possibilities of directional movements of parts of the trunk.*

> You will discover very exciting features which are governed by some of the following factors:
> *(a)* small parts of the spine can move into different directions, however, without changing levels;
> *(b)* the two ends of the spine — the top (including or excluding the head) and the pelvis area — can also move into different directions.
> All parts of the spine can move:
> *(i)* separately as other parts are still; or
> *(ii)* in concert with other parts, each moving in different ways.

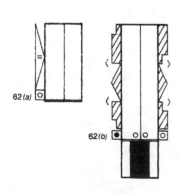

62(a)

62(b)

Principles of Trunk Movement

If we now sort out the result of the various actions which we have found in the two last exercises we might group them under the following headings.

  (a) Radius-like, moving at an angle from the hip joints.
  (b) Pincer-like, curling from one or both ends of the trunk.
  (c) Bulge-like, shifting the central area of the trunk out of its normal position.

Combinations of these fundamental principles can be observed in the manifold actions of the trunk — both in stillness and in motion — often occurring in a very narrow extension of space and with subtle use of energy and time differentiations. It will be useful to give further consideration to the upper and the lower parts of the trunk, as each is collaborating with a different set of limbs.

Since all our movements, but particularly the carriage of our body, are influenced by the physical law of gravity we might refer in this connection also to the "centre of gravity" which in the human body is situated in the pelvis region and is, in the normal mode or carriage, above the point of support.

The upright carriage of the human being brings into prominence the actions of the chest and in this situation we are particularly aware of the breast-bone to which we might refer as the "centre of levity".

The shoulders can move around the spine and the centre of levity quite independently and also independently from one another.

63. *Perform a quivering movement with the shoulders, i.e. a series of quick actions, not involving the trunk otherwise, and come to a halt which is tense and produces an asymmetric carriage of the body. (Asymmetric means that the directional placements of the shoulders differ from one another.) After a pause bring your shoulders slowly back to their normal symmetrical position, carrying the head high. Repeat several times.*

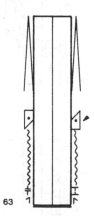

63

64. *Perform the same movement sequence as Exercise 63, involving a definite action of the:*
    (a) upper part of the trunk;
    (b) lower part of the trunk;
    (c) whole of the trunk.

The Chest "Looking"

The observation and analysis of bodily actions will be very much simplified if we understand that those of some parts of the body are done, consciously or unconsciously, in relation to an outer object or point of interest.

This fact becomes most evident in the movements of the head which frequently result from directing our face towards something at which we wish to look. In this way we may also refer to the palms of our hands, the soles of our feet as "looking" in the direction of an object which is to be approached, touched, gripped, kicked.

Similarly, we can ascribe to the front of the chest, with the breast-bone in its centre, the function of "looking". This facilitates the analysis of those actions of the trunk leading to a twist.

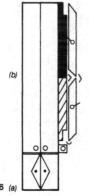

65

65. *Stand with both legs apart and "look" with your breast-bone once to right-forward and once to left-forward. Take care that the lower part of trunk or pelvis region remains in its original position.*

66. *Perform the same movement as in Exercise 65, but while turning chest to:*
    (a) right-forward the upper trunk, including head, moves backward-high;
    (b) left-forward the upper trunk, including head, moves forward-deep.
    Repeat this action several times and let the legs and arms assist in making the movement a fluent one.

66 (a)

The Pelvis "Pointing"

The lower part of the trunk is much less mobile than the chest and will therefore usually follow the movements of the hips which, like all the other articulations, "point" or move into a certain direction.

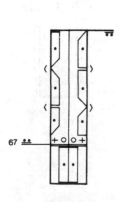

67

67. *Keep feet together, whip right hip to right-forward while left hip goes left-backward; then left hip to left-forward and right hip right-backward. Do this several times in quick succession, ending with right hip right-forward.*
    Repeat, starting with left hip left-forward. Take care that the chest stays in its original position.

68. *Perform same actions of hips as in Exercise 67, but as the right hip comes forward for the first time, the upper trunk, including head, is thrust forward-upward. It should pause in this position for the remainder of the sequence. The upper trunk is thrust backward-upward with the first movement of the repeat, and should again pause to the end of the sequence.*

69. *Invent a sequence with rhythmical contrasts combining the actions of the trunk, as outlined in Exercises 65 and 67.*

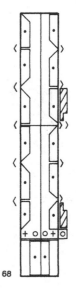

68

Since the breast-bone can be considered as "looking" into directions such as forward-down or forward-up, it will be an interesting study to find out the differentiation between these actions and those referred to as movements of the upper trunk going, or "pointing", into a certain direction.

At any rate, in all trunk movements one has to become keenly aware which area of the spine, of the chest and of the abdomen is:

(a) involved in the action;

(b) pausing in its original position;

(c) carried along passively, as for instance in the shiftings of the chest when the whole of the upper trunk is displaced from its natural position above the pelvis.

Other Parts of the Body "Looking"

While hands, feet and head have so far been regarded as part of the actions of the limbs and the trunk respectively, they may take on an orientation of their own.

As already mentioned, in such a case the palms, the soles and the face become the areas which are related to a direction of space towards which they "look".

70. *Experiment with one of your palms looking:*

downwards;

upwards;

forwards;

backwards;

to the right;

to the left;

or in any other direction.

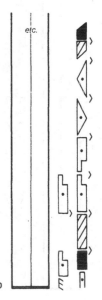

70

71. *The same as Exercise 70 for the sole of one foot.*

> To do this exercise will involve balancing the body in different situations.

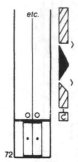

72. *Look with your face into certain directions.*

> You will notice that in many instances the trunk has to help in order to make this possible, for instance, when looking backward.
>
> The "looking" palms, soles, face and breast-bone cause a lot of twisting movements of the arms, legs and trunk. The mastery of bodily actions is greatly helped by acute awareness of the limitations in the degree of twist as well as of the possibilities of its extension by involving neighbouring areas of the body.

Turns of the "Looking" Parts of the Body

Besides "looking", palms, soles, face and breast-bone can also be turned. The axis around which they turn is related to the looking direction. In the case of head and chest, the pivoting point is the centre of the area moved, while feet would pivot around ankles, and hands around wrists (or even elbows). This idea may require at first a little adjustment by the reader as certain movements which are commonly referred to as bendings will in fact be turnings. For instance:

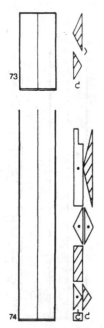

73. *The face looks forward and turns alternatively clockwise and anti-clockwise around the forward direction as far as it will go.*

> This means that the right ear alternates with the left ear in approaching right or left shoulder respectively.

74. *Face looking to right side with clockwise turn (right ear approaches back).*
Face looking up, retaining the clockwise turn, even increasing it according to the possibility of the new position (this brings the right ear to forward).
Face looking to left side, still retaining the clockwise turn as far as possible (the right ear approaches breast-bone).
While face is gradually moved forward the head is turned in an anti-clockwise manner (left ear approaches left shoulder).

75. *Experiment similarly with your palms starting in the*
    following way:
        both arms are narrow forward
        right palm looks to left side
        left palm looks to right side
        both hands are turned clockwise
        (this will bring the fingertips of left hand near
        breast-bone and those of right hand away from it).

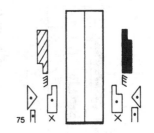

While prominence of expression is often given to the
hands, feet and head in that they seem to "look" into this
or that direction, at other times they just follow passively
the movements of that part of the body to which they
belong. In such cases the hands take part in the actions of
the forearm or arm, the feet in those of the legs, and the
head in those of the chest or trunk, and they have no
orientation of their own.

Twists of the "Pointing" Parts of the Body

In addition to the twists of the trunk (which have already been discussed
in connection with Exercises 65 to 69), twists may also occur in the limbs
independently from the influence of "looking" hands and feet.

It will be useful to regard these as rotations:

(a) inward when thumb or big toe side respectively of the limbs twist
    towards the centre line of the body;
(b) outward when they twist away from it.

This is best started from the normal position of these limbs.

Such twists can be observed, for instance, in gestures of arms and legs
when these perform continuous movement in the shape of a figure 8, or
an S curve, as for instance in Exercise 18.

76. *Perform Exercise 18 with obvious twisting move-*
    ments of the arms, combining the starting position of
    both arms with an inward rotation. Rotate outward
    while moving via forward to deep. Follow this by an
    inward rotation together with a raising of both arms
    to either side.
    Do this sequence in a successive way, starting with
    shoulders and ending with hands, making sure that
    the movement continues fluently and supplely.

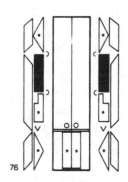

Peculiarity of Arm Twists

Twisting of arms can be done both in a successive and in a simultaneous
way, and the rotation can be started either from the shoulder or from the

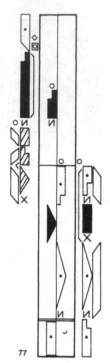

hand. When moving successively the whole arm is only gradually involved in the twist, while in simultaneous action the twist occurs in the whole arm at once.

> 77. *Explore various possibilities of twists in arm gestures, using different extensions of space and creating definite shapes. In this connection look once more at Exercises 22 and 25.*

Peculiarity of Leg Twists

Twists in the legs can only be done in a simultaneous manner, which means that the leg as a whole is rotated inward or outward. Only the feet can have an isolated rotation. When using the legs for support, this leads to various possibilities in the style of standing and stepping.

Positions of the Feet

When the weight is carried by both legs a distinction is usually made between closed and open positions, closed being those when the legs are together and open when they are apart.

In each category various forms can be observed which are based on different spatial relationships of the legs, as well as on the fact that the legs may be rotated to some degree.

It is assumed that the reader knows the five basic positions of the feet which in the classical ballet style are performed with extreme outward rotation of the legs while in the contemporary dance and in many national dance styles the outward rotation is less pronounced.

First, third and fifth positions are closed positions. Each in turn gives an increasing sensation and appearance of the legs being knit together as one. This is because from a heel to heel position a heel to toes position is achieved by crossing one leg in front of the other.

Second and fourth positions are both open positions, one stressing the right-left symmetry of the body and giving it stable support. The other, a forward-backward stance, represents a more alert position from which locomotion can be initiated.

In an open position the centre of gravity is above a point on the ground between the two legs. If one of the legs is lifted it is moved to a position above the other leg.

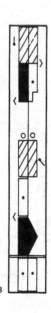

> 78. *Invent step sequences in which pauses in different positions occur.*

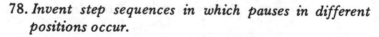

> Observe that you have to shift your centre of gravity after pausing on both feet and before performing the next step.

79. *Combine gestures, steps and positions using different kinds of rotations in the legs.*

> Note that a number of open positions can be established other than the two aforementioned by using different kinds of spatial relationship between the legs, e.g. in a diagonal, or a sideways crossing direction.
> Open and closed positions are also possible in which the weight is carried on one leg only while the other might touch the floor.

80. *Explore different ways in which you land on one foot and the other touches the floor after jumping. Arrive similarly after leg gestures, steps or turns leading from, via, and into open or closed positions.*

> Make sure that you sometimes use inward and outward rotations of the legs, even "looking" and "turning" feet.

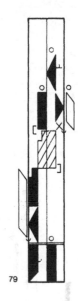

79

Whole and Half Steps

It might be mentioned here that we can distinguish whole and half steps.

Our ordinary way of walking consists of whole steps, i.e. each leg in turn passes the centre line of the body when moving from backward to forward.

A whole step can be cut into two halves.

81. *Stand on right foot, the left is backward-deep touching the floor.*
    Step: left foot to stance medium, then
    right foot forward-medium.
    Repeat a number of times.

> Notice how this sequence of half steps creates a certain feel of control and hindrance, while a sequence of whole steps gives an expression of freedom.
> Many dance steps are combinations of these two, giving rich possibilities for expression in travelling movements. Look at Exercises 4 and 5 which give two basic step-patterns, Exercise 4 consisting of: half, whole, half step; Exercise 5 consisting of: whole, half, half step.

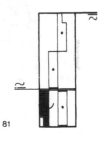

81

82. *Stand on left foot, with the right leg rightside-deep touching floor with big toe. Slip swiftly right foot into a closed position, knocking the left foot away from its stance and transfer weight to right leg.*
    Left leg has meanwhile moved to leftside-deep, touching floor with big toe.

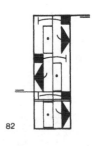

82

Repeat action with other side, and then alternate sides, taking care that the centre of gravity is all the time held over the same spot.

Movements of the Centre of Gravity

As has been stated, in the normal upright carriage of the human being the centre of gravity is situated above the point of support.

In fact, when the body is in a stable balance the centre of gravity is placed in a vertical line with the point of support. When standing on feet or hands it is above the support, when crouching it is at the same level with it and when hanging, for example from a bar, it is below.

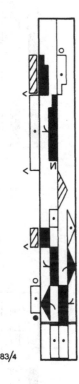

83. *Move the centre of gravity between medium level (i.e. when seat is near the supporting feet) and high level as you make wide and narrow steps into different directions. Introduce to these sequences pauses in open or closed positions.*

> Note that deep steps, i.e. steps with bent knees as explained in connection with Exercise 5, as well as high steps cause only slight changes of level of the centre of gravity. These changes are part of the nature of the steps and not deliberate movements of the centre of gravity. The same applies to the normal elevation off the ground and also to hand-stands when the shoulders are lowered towards the ground.

84. *Perform Exercise 82 with centre of gravity in medium level, that is, near the supporting foot. This movement is like a Russian step in a crouching position.*

83/4

Unstable Equilibrium

In our considerations of bodily actions we have, up to now, mainly worked with stable balance. In the mastery of movement the unstable or labile state of equilibrium plays an important part. This is when the centre of gravity is apt to change its normal vertical relation to the point of support.

Special co-ordinations of bodily actions can often be observed in the human being in order to counteract loss of balance, or to re-establish balance.

Two main ways of bringing about loss of balance can be distinguished:
(a) the centre of gravity is moved into a direction in space while the supporting part of the body has no action; or
(b) the support of the body is removed without the centre of gravity being shifted into any direction in space.

If the body does not undertake any suitable counteraction a fall will result in both cases.

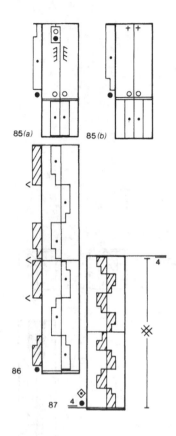

85. *Try to execute some falls initiating these through movements of the centre of gravity.*
    (a) Forwards; here you will fall flat on to your whole front.
    (b) Backwards; this will result in a sitting position.

86. *Do the same as in Exercise 85 (a) and (b) but prevent the fall by a number of steps.*

    This will give the steps a particular character of a precipitating nature.

87. *Perform a few very small and quick steps forward without shifting the centre of gravity; when you are just on the brink of losing balance, quickly reverse the steps to backward (behind the line of gravity).*

    Repeat this sequence several times until the sensation is felt of having the legs swept away from underneath.

88. *Invent your own sequences of actions, freely alternating between stable and labile equilibrium involving different levels of the centre of gravity.*

    Observe the effect this has on the control of the flow of movement and on the kind of energy used.

89. *Experiment with flying and falling movements by springing free from the ground and with your body leaning obliquely. Try to land in such a way that the centre of gravity is not vertically above your feet.*

    If you are not counteracting this unstableness in any way, you will fall on the ground and probably roll before you come to rest.
    The body is in its most *labile* state of action when the centre of gravity is propelled diagonally upward or falls in a diagonally downward direction and is out of line with its normal support.
    The body is in its most *stable* state of action when the centre of gravity is in vertical line above the point of support.

Supporting Weight on Various Parts of the Body

The bodily actions involved in supporting and transferring weight on the various parts of the body, such as knees, hips, trunk, shoulders, head, elbows and hands, will be dealt with here only in so far as this may add important new aspects of observation and analysis.*

The main situations, besides standing, in which the body pauses or comes to rest are:

kneeling — sitting — lying.

Kneeling and Sitting Positions

Between the positions of upwards in standing and horizontal in lying we find those of kneeling and sitting which may act as transitional stages during a gradual change between the two extreme situations. They may, of course, occur also as positions in their own right, each having its own characteristic range of bodily expression.

90(a)   90 (b)

90. *Explore the possibilities of kneeling on both knees. Notice the spatial relationship of the knees:*
    (a) in closed positions;
    (b) in open positions.

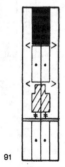

91. *Kneel "high", "medium", or "deep", i.e. hips above knees, hips halfway down, hips on heels.*

91

92. *Explore which directions you can use when transferring weight from a standing position on to one or both knees:*
    (a) at the same time;
    (b) one after the other.

93

93. *Explore whether there are any directional limitations in the transference of weight on to one or two hips from a kneeling position.*

Note that directions backward, or half backward, lead to sitting positions while all others produce forms of lying positions. In order to get into these from kneeling some weight may have first to be transferred on to a hand before being able to proceed with the hip action.

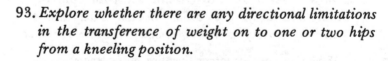

* They are, together with others, explained in great detail in Albrecht Knust's *Dictionary of Kinetography Laban (Labanotation)*, Macdonald & Evans, 1979.

Lying Positions

In these the weight is supported by the whole torso, involving hip and shoulder regions alike.

94. *Perform actions of the body which will lead to different lying positions. Use various spatial directions, particularly for the transference of weight on to the shoulder region. These have to be related to the preceding support on the hips.*

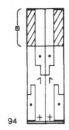

94

Note that only when transferring weight on to the hips and shoulders in a straight-forward or a straight-backward direction do we lie flat on the ground; in the first case on the abdomen, and in the second on the back.

95. *Start in a kneeling position and transfer weight on to the torso or parts of it while pausing with the knees, this means while remaining on them. Certain twists in the trunk may result.*

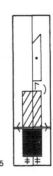

95

It is interesting to observe the various forms of carriage which the body can assume in standing, kneeling, sitting and lying situations.

These are greatly influenced by structural and functional factors of the body, such as:
  (a) the spine and its *pin-like* extension;
  (b) right-left symmetry of the body and its *wall-like* surface;
  (c) the limbs, together with their respective trunk regions, curling and circling in *ball-like* shapes;
  (d) shoulder-girdle and pelvis twisted against one another in a *screw-like* fashion.
In your experiments of weight supports in various situations become aware of the particular body carriage and its shape.

When lying on the floor or when in an upside down position it is not always easy to find one's directional bearings. It is as well to remember that gravity always acts vertically to the earth and that our directional sense is related to this, i.e. up always remains up, no matter whether the legs or the head are up.

For instance, when lying on your right hip and shoulder, your chest is looking forward, your legs are extended to left side, and your left arm might be lifted upward.

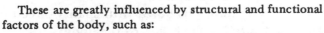

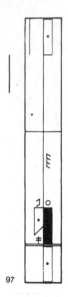

97

96. *Invent transferences of weight on to various parts of the body:*
    (a) simultaneously;
    (b) successively.

> Take care that with each action the weight is fully transferred, i.e. the weight-bearing body part(s) at one time will next time be released off the floor.

97. *Start standing on one foot and perform 96 (b), but after each transference the support is retained (pausing). In this way, several body parts together with the original foot will finally carry the weight. Re-establish gradually the initial position by performing gestures with one after the other of the weight-bearing parts by releasing their support.*

> Observe that the centre of gravity must be lifted some time, otherwise you will remain crouching on one leg.

Forms of Locomotion

There are many possibilities of locomotion by using body parts other than feet, or in addition to them. The following points give a rough survey of how progression through space can be made:

    (*a*) by continuous contact with the floor in a directional manner;
    (*b*) by releasing contact with the floor and moving into various directions free from the ground; and
    (*c*) by some form of turns combined with spatial directions. The rotations are related to the three axes of the body;
       (i) head to foot,
     (ii) front to back,
   (iii) side to side.

98. *Investigate the following as to their axis in rotation and to their direction in progression through space:*
    cartwheeling;
    somersaulting;
    rolling.

98

Floor Patterns

Floor patterns are produced through locomotion either:

    (*a*) by deliberate orientation in the various forms of locomotion; or
    (*b*) by incidental shifts of the weight resulting from the various actions of the body, particularly of those which stimulate progression through space, such as wide extensions and labile movements.

Mastery of movement requires clarity of pattern in either case. What kind of designs are created will depend on the use of:

(a) the particular spatial directions with or without change of front by means of a pivot-like turn; or

(b) the particular spatial directions together with change of front taking place gradually during several transferences of weight, say a number of steps.

99. *Strut, leap, crawl, dash or sneak along, using the same direction for each transference of weight on to some body part or parts.*

Sequences of transferences of weight into one direction form paths on *straight* lines.

100. *Do the same as in Exercise 99 but use different directions.*

Sequences of transferences of weight with abruptly changing directions form *angular* paths.

101. *Produce a Z-like floor pattern with a number of steps.*
(a) Without change of front.
   Notice the directions of your steps in each phase of the movement.
(b) With change of front, stepping all the time to the left side.

Notice: (i) where on the floor the pattern will appear in relation to your starting position, and (ii) how far you have to pivot at each corner in order to continue with steps to the left.

102. *Step into the following directions without change of front:*
forward-right, rightside, right-backward, backward, backward-left, leftside, left-forward, forward, forward-right.

You will have completed a *curve*, created a circle. Notice that the directional value of each step shows only a slight change from each previous one.

Curved paths may also come about by a gradual change of front while retaining the same stepping direction.

Look once more at Exercise 52 in which you performed half a circle by stepping backwards. The centre of this is situated on your right side.

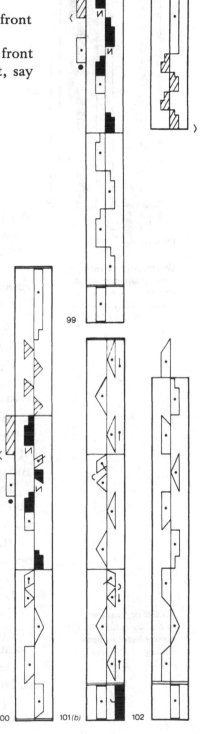

99

100   101(b)   102

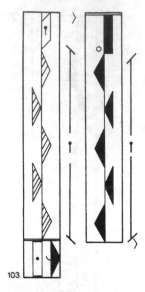

103

Changing relationships, created when moving, can result in physical contact; the particular role of face and hands in this

Spatial organisation of bodily actions in the three phases of contacting

103. *Produce an S-curve-like floor pattern; this consists of a double bent curve with two different centres. Keep your front all the time at first towards one and then towards the other centre.*

Start in such a way that during the gradual change of front in the first curve you step to your right side, and in the second curve you perform steps to your left. You must, however, insert a change of front (half turn) between the two curves, in order to do this and to face the centre of each.

MOVEMENTS RELATED TO PEOPLE AND OBJECTS

When moving we create changing relationships with something. This something can be an object, a person or even parts of our own body, and physical contact can be established with any of these.

In doing this we can distinguish three main phases which are brought about by appropriate bodily actions:

(a) preparation;

(b) actual contact;

(c) release.

Face and hands play a particular role in this. The face as the main seat of our senses will come into prominence mainly in the preparatory phase, while the hands, which are the natural instruments of gripping, will come to the fore in the one of actual contact.

One can clearly observe when eyes or ears are directed towards a point of interest, or when the nostrils widen to take in a real or imagined scent and when the tongue licks the lips in anticipation of a taste.

The hands of the human being are wonderful instruments which can perform most complex movements. Although it may be possible to reduce these to their fundamental function of gripping and repulsing, the actions of the single fingers in relation to one another as well as to objects and persons lead to a great variety of situations which are particularly relevant to the second phase.

Let us consider the three phases in more detail.

To the preparatory phase belong such actions as "looking" of face, hands, feet, chest. This results in what we might call "addressing" the point of interest which can happen from one direction.

Next we might "approach" and "meet", or "surround", or "penetrate" it. No contact is yet made and any spatial directions can be used.

Approach, involving either locomotion or gesture, or both, could be made from any side of the object in question.

In meeting, the spatial relation in mutual placement is important, and it has any directional freedom (of course, within the limits of the situation). Surrounding requires several parts of the body moving around the point of interest from several directions, while penetrating means going into it from one direction.

The actual contact can be brought about by:

touching	from any direction;
gliding	along any direction of object surface;
transferring weight	from above object;
carrying	from below object which rests on a body part;
holding	on to any side of the object, by surrounding it with several body parts.

Bodily actions which cause release can go anywhere as long as they lead into a direction contrary to that of the object contacted.

In relating our movements to people or objects there is, of course, no need for physical contact. The climax of such movements is then to be found in the moments of meeting, in confronting, passing, jumping over or surrounding, crawling under or diving, embracing, weaving through, and suchlike. *Climax of relationship without contact*

While both partners and objects can be stationary, they can also be mobile.

Moving objects, for instance a bouncing ball or a rolling hoop, can also be observed travelling along a definite path in space. Pieces of clothing, veils, mantles, etc., which are put on and taken off could be looked upon as such moving objects; ropes, revolving stages, screens or any other theatrical machinery which is set in motion can also be described in their spatial pattern in relation to people and things. *Spatial patterns of moving objects and people*

While such measurable factors as speed, force, direction and extension are common to both moving objects and moving people it is quite clear that the movements of the human body differ vastly from those of machines. Even when man sets about a working job and his bodily actions have to fulfil practical functions they are distinguished by personal expression. Sometimes, however, they may be rather devoid of inner participation and then become mechanical. Yet, sometimes they are richly shadowed by effort patterns serving no practical purpose at all. Since these spring from the very root of personality, they create characteristic expression which becomes visible in bodily action. *Bodily actions have personal expression and are shadowed by effort patterns*

CORRELATION OF BODILY ACTIONS AND EFFORT

As previously mentioned, effort is manifested in bodily actions through Weight, Time, Space, Flow elements. Not all of these motion factors are always significant and according to their combination they produce particular shadings.

Characteristics of the action drive

In the above deliberations, Weight, Time, Space were almost exclusively used. That was because a particular movement drive of the human being was studied, namely that of Action. This action drive is characterised by performing a function which has concrete effect in space and time through the use of muscular energy or force.

In a living being such actions are never devoid of expressive elements, which means that they cannot be determined by logical reasoning, nor grasped by measurable factors only. They are pervaded by elements which bring out the quality and the attributes of particular species.

While animals' movements are instinctive, man can become conscious of his effort patterns and reshape them

While animals' movements are instinctive and mainly done in response to external stimuli, those of man are charged with human qualities and he expresses himself and communicates through his movements something of his inner being. He has the faculty of becoming aware of the patterns which his effort impulses create and of learning to develop, to re-shape and to use them.

The actor, the dancer, the mime, whose job it is to convey thoughts, feelings and experiences through bodily actions, has not only to master these patterns but also to understand their significance. In this way imagination is enriched and expression developed.

Introduction to basic terms of a movement analysis

The following sequence of movements will serve as an introduction to the basic terms of a movement analysis necessary for the deeper understanding of movement which is not only the actor-dancer's material but a phenomenon fundamental to life.

In an inert human body the only motion observable is the result of respiration. The alternate rising and sinking of the weight of the chest occurs in space. In sighing we could also hear this respiration. Both the rising and falling of the chest and the sounds of breathing have a certain duration in time. The whole motion observed is in continuous flow; there is no noticeable stop. If one feels the pulse, or puts one's ear to the chest of the recumbent body, another continuous inner motion, the heart-beat, is detected. In rising, the body shows an increased state of motion. The breathing was a light, relaxed movement, but now, more weight being lifted, there is a strong tension of muscles as the whole body is lifted and stretched.

The effort made in getting up is, however, not only composed of different light and strong muscle tensions, but also varied in space form. In the first movement of rising the body becomes bent and twisted, and is therefore flexible in all its parts; the second stretching movement is direct. Flexible movements take many directions in space simultaneously, but the stretching-up movement a few only, and, finally, the definite direction upwards. The duration of the act of rising can be short or long.

It can be a sudden jumping up, or a sustained movement into the upright position. The quick movement takes no more time than one breath or heart-beat; the sustained lifting of the body takes the time of many such breaths or heart-beats. The flow of the whole movement can be fluent like regular respiration, or interrupted as when one is out of breath.

Weight, Space, Time, and Flow are the motion factors towards which the moving person adopts a definite attitude. These attitudes can be described as: *Attitudes towards the motion factors*

  a relaxed or forceful attitude towards weight;
  a pliant or lineal attitude towards space;
  a prolonging or shortening attitude towards time;
  a liberating or witholding attitude towards flow.

A specific combination of several of these eight elements of movement is observable in every action, and is more evident in the so-called basic actions* in which the easily discernible factors of Space, Time and Weight are mainly considered. *Introducing "basic actions"*

In a quick reaction to some external stimulus four different actions are mainly used:

  a direct and strong thrust;
  a flexible and strong slash;
  a direct and light dab;
  a flexible and light flick.

By outer resistance or inner hesitation the action can be delayed and become sustained in the following four actions:

  a direct and strong press;
  a flexible and strong wring.
Or, if the resistance is less strong:
  a direct and light glide;
  a flexible and light float.

Certain movements can be considered derivatives of basic actions. *Derivatives of basic actions*

Basic Action	Derivatives
thrust	shove, punch, poke
slash	beat, throw, whip
dab	pat, tap, shake
flick	flip, flap, jerk
press	crush, cut, squeeze
wring	pull, pluck, stretch
glide	smooth, smear, smudge
float	strew, stir, stroke

*See also page 170.

Actions in which one or two of the elements of Time, Space or Weight are almost entirely absent, or slightly stressed, are incomplete elemental actions.

In casual movements that are neither sustained nor quick the time element has no physical import. Such movements seem to consist of Weight and Space components only which means that there exists no definite attitude of the moving person towards the time factor. They are incomplete elemental actions which can frequently be observed as transitions between basic actions. For example, a person lifting a heavy sack on to his shoulder will start with a kind of slashing movement which develops into a wringing movement. Between the quick slashing and the sustained wringing, a transitional phase can be observed, during which the two effort elements common to slashing and wringing, namely strength and flexibility, are maintained, while any stress on time quality disappears, which means that neither quickness nor sustainment is felt. Wringing developing into slashing, as observed in energetic tearing, produces the same intermediary effort. We describe such intermediary movements as lying between two basic actions. Applied to the previous example, the intermediary actions will fall "between wring and slash" or be shortly described as "wring-slash".

The other incomplete actions arising frequently in everyday life are as follows.

Dab-glide is a transitional effort arising when dabbing develops into gliding, as, for instance, when brushing away a light object. The same intermediary effort might appear when gliding develops into dabbing, say, in shutting a small drawer or inserting a card into a file.

Flick-float will accompany small wheeling movements, as in twiddling one's fingers or in throwing small things away.

Thrust-press, where thrusting develops into pressing, can often be observed in assembling or modelling kinds of movement.

These incomplete actions can not only appear as steps between two basic actions but can also be produced as cardinal movements. For instance, small movements accompanying speech are often incomplete efforts in which the attitude towards one or the other motion factor is neutral. If someone says "Oh, never mind", the actor might accompany his words with a small gesture which will probably be round (flexible) and quick. Now, it must be said that there exist two forms of flexible and quick movements, one of them slashing, which is strong, and the other flicking, which is light. The small gesture accompanying the words "Oh never mind" will rarely be a strong slash or light flick, but between these two movements. It can be described as "between slash-flick", or shortly as *slash-flick*, which means that the weight element remains unimportant or, expressed technically, unstressed.

Let us briefly look at the eight basic effort actions* and then the incomplete elemental actions and movement drives. We shall recognise in them measurable components such as have been previously surveyed at length. Now let us at the same time direct our attention to the vital or organic elements which are always inherent in any effort display, no matter whether it serves practical purposes or is of an expressive nature.

Basic Effort Actions

The eight basic effort actions represent an order of Weight, Time, Space combinations which is built up on two main mental attitudes, involving on the one hand objective function and the other movement sensation. While neither is ever entirely without the other, owing to the unity of effort, each can become so dominant that for practical reasons it can be referred to on its own. The two attitudes have been previously referred to as "fighting" or "resisting" and as "indulging" or "yielding".

The polarity of mental attitude towards the motion factors of W.T.S. results, when combined, in eight characteristic effort actions

The basic action springing from an attitude of fighting is:

The two basic actions which combine sameness of attitude towards W.T.S.

firm, sudden and direct which we meet in movements of *thrusting*, punching, stabbing, piercing, etc.

The basic action completely unrelated to this, springing from an attitude of indulging, is:

gentle or fine touch, sustained and flexible, which we meet in movements of *floating*, slightly stirring, flying, drifting, rousing, etc.

These two effort actions each form a centre to which three of the other six are closely related. We can see that by taking away one element at a time and replacing it by a foreign one, that is, one originating from the other attitude, the original action is mutated into another basic effort.

Mixing opposite attitudes produces six modifications of the original two basic actions

* These are explained in detail in *Modern Educational Dance* by R. Laban (third edition 1975) and *Effort* by R. Laban and F. C. Lawrence (second edition 1974), both published by Macdonald & Evans.

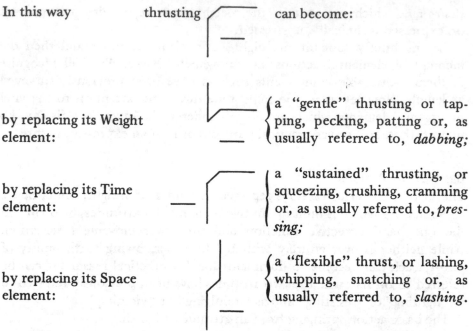

In this way   thrusting   can become:

by replacing its Weight element: { a "gentle" thrusting or tapping, pecking, patting or, as usually referred to, *dabbing;*

by replacing its Time element: { a "sustained" thrusting, or squeezing, crushing, cramming or, as usually referred to, *pressing;*

by replacing its Space element: { a "flexible" thrust, or lashing, whipping, snatching or, as usually referred to, *slashing.*

Dabbing, pressing, slashing form one group of basic effort actions which is closely related to thrusting (*see* (A) below).

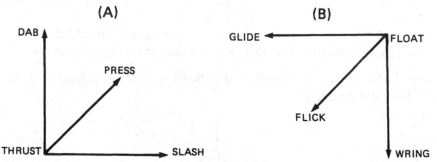

(A)   (B)

The other group, wringing, flicking, gliding, is closely related to floating or stirring (*see* (B)) and develops as follows:

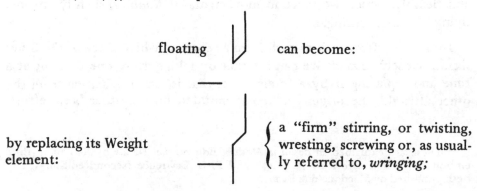

floating   can become:

by replacing its Weight element: { a "firm" stirring, or twisting, wresting, screwing or, as usually referred to, *wringing;*

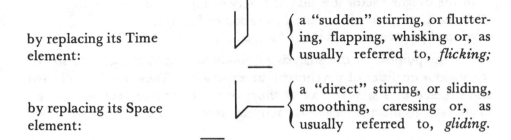

by replacing its Time element:

{ a "sudden" stirring, or fluttering, flapping, whisking or, as usually referred to, *flicking;*

by replacing its Space element:

{ a "direct" stirring, or sliding, smoothing, caressing or, as usually referred to, *gliding.*

Analysis of Effort Elements

In an attempt to analyse the various effort elements we might define them in the following way. There are two components in each, one of which is operative and objectively measurable, and the other, personal and classifiable.

Measurable and classifiable components

Weight: The effort element "firm"
consists of *strong* resistance to weight, and of a movement sensation, *heavy*, or a feel of weightiness.
The effort element "fine touch" or "gentle"
consists of *weak* resistance to weight and of a movement sensation, *light*, or a feel of weightlessness.

Time: The effort element "sudden"
consists of *quick* speed and of a movement sensation, of a *short* span of time, or a feel of momentariness.
The effort element "sustained"
consists of *slow* speed and of a movement sensation of a *long* span of time, or a feel of endlessness.

Space: The effort element "direct"
consists of a *straight* line in direction and of a movement sensation of *threadlike* extent in space, or a feel of narrowness.
The effort element "flexible"
consists of a *wavy* line in direction and of a movement sensation of *pliant* extent in space, or a feel of everywhereness.

While in functional actions the movement sensation is an accompanying factor only, this becomes more prominent in expressive situations where psychosomatic experience is of utmost importance. In such situations we can observe changes of emphasis within the Weight, Time and Space factors of bodily actions. A set of aspects other than Resistance, Speed and Direction becomes apparent, which is particularly relevant to movement sensations. Here the pertinent aspects are as follows:

Psychosomatic experiences in expressive movement

of the Weight Factor it is that of Levity or Lightness;
of the Time Factor it is that of Duration or Passage;
of the Space Factor it is that of Expansion or Plasticity.

For the purpose of the analysis of bodily actions it is useful to survey the various qualities of psychosomatic experience. These can be collected and ordered in a similar way as those of the eight basic actions and we might refer to them as eight basic movement sensations.

Movement Sensations

The two fundamental ones are:

(a)   this as an effort action is floating and its inherent movement sensation is what might be called *suspended*.

In such a movement the psychosomatic experience:

> of levity is *light* as if buoyed aloft;
> of duration is *long* as if existing in everlasting time;
> of expansion is *pliant* as if crumpled in space.

(b)   this as an effort action is thrusting, and its inherent movement sensation is what might be called *dropping*.

Here the mentioned psychosomatic experiences are all diminished:

> that of levity becomes *heavy* as if dragged down;
> that of duration becomes *short* as if existing in a single moment of time;
> that of expansion becomes *threadlike* as if guided through an extended and uniform aperture in space.

The three basic movement sensations closely related to "suspended" may be described as follows.

Relaxing.   In this the Weight has changed and become *heavy*, therefore its three components are:

> heavy — pliant — long.

Excited.   In this the Time sensation has changed and become *short*, therefore its three components are:

> light — pliant — short.

Elated.   In this the Space sensation has changed and become *threadlike*, therefore its three components are:

light —threadlike — long.

The other three which are closely related to "dropping" might be referred to as follows.

Stimulated. Here the Weight sensation has changed and become *light*, therefore its three components are:

light — threadlike — short.

Sinking.   Here the Time sensation has changed and become *long*, therefore its three components are:

heavy — threadlike — long.

Collapsing. Here the Space sensation has changed and become *pliant*, therefore its three components are:

heavy — pliant — short.

Movement sensations giving psychosomatic experiences can be observed in the bodily actions. They have no objectively measurable properties and can only be classified with regard to their quality, their intensity and their rhythm of development. They are states or moods giving particular colour to bodily actions.

The Motion Factor of Flow

Let us now consider the motion factor of Flow* in more detail. It plays an important part in all movement expression, as through its inward and outward streaming it establishes relationship and communication. It is mainly concerned with the degree of liberation produced in movement no matter whether this is considered from the point of view of its subjective-objective opposites or the contrasts of being "free in" or "free from" the flow of movement. A description of flow involves its complete negation, which is stop or pause. It involves also the motion of resistance and counter-movement, each of which is different in mood and meaning and is not to be taken to refer to direction, speed or strength.

Flow, the motion factor which establishes relationship and communication

Previously we considered the basic property of movement, namely its natural flux. We observed the fact that this in its extremes can be either going on or completely stopping. Now in an attempt to define the motion factor of Flow with its effort elements "free" and "bound" we have also to take into consideration the movement sensation of Fluency.

This is concerned with the ease of change as may be observed in the movement of a fluid substance. When the sensation of flowing on is

See also page 172.

reduced one could eventually speak of "pausing" in which, although still, we have the feel of a continuation but of a withheld one.

> The effort element of "bound" or hampered flow
> consists of the readiness to stop normal flux
> and of the movement sensation of *pausing*.

This may be a somewhat difficult conception. It might help, however, if we realise that the sensation of fluency, the feel of being carried on, does not cease when pausing, but is controlled to the utmost.

Bound flow then, when combined with other effort elements, say with strong and direct, gives that movement a quality of restraint. It originates from an inner preparedness to stop the action at any given moment. The flow seems to stream backward towards the central area of the body and in a contrary direction to that of the action.

> The effort element of "free" flow
> consists of released flux
> and of the movement sensation of *fluid*.

The effort quality of free flow should not, however, be confused with the movement sensation of merely going on. This sensation is in effort expression activated by the released flux. With its emitting capacity, it helps the progression of movement through the body from the central area towards the extremities thus producing a feeling of onward streaming which is characteristic of free flow.

Table VI
EFFORT
*Survey of the Aspects of Weight, Time, Space and Flow Needed for the
Understanding of Effort*

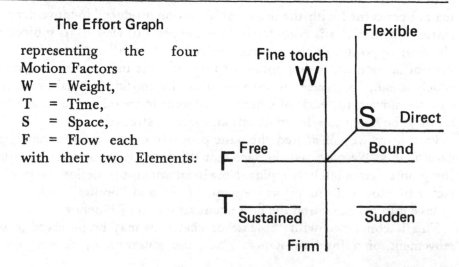

The Effort Graph

representing the four Motion Factors
W = Weight,
T = Time,
S = Space,
F = Flow each
with their two Elements:

Fine touch / Flexible / W / S Direct / Free / Bound / F / T / Sustained / Sudden / Firm

Motion Factors	Effort Elements		Measurable Aspects (objective function)	Classifiable Aspects (movement sensation)
	(fighting)	(yielding)		

Weight	firm	gentle	Resistance
			   strong (or lesser
			   degrees to weak)
				Levity
				   light (or lesser
				   degrees to heavy)

Time	sudden	sustained	Speed
			   quick (or lesser
			   degrees to slow)
				Duration
				   long (or lesser
				   degrees to short)

Space	direct	flexible	Direction
			   straight (or lesser
			   degrees to wavy)
				Expansion
				   pliant (or lesser de-
				   grees to theadlike)

Flow	bound	free	Control
			   stopping (or lesser
			   degrees to releas-
			   ing)
				Fluency
				   fluid (or lesser
				   degrees to pausing)

It will now be possible to tabulate the four motion factors and their effort elements and to list the aspects contained in each factor; this is done in Table VI above.

Incomplete Elemental Actions

As we have previously seen, in studying bodily actions we can often make the strange observation that one of the motion factors is entirely neglected and only two, of which flow may be one, seem to give the shading to the movement. We speak in such cases of "incomplete effort". Equally often, we can see that the motion factor of Flow has taken the place of one of the three others which remains latent, still leaving three active. In such a case the bodily actions have quite a different quality and spring from movement drives different from that of Action. This means that the impelling forces leading to activity may be composed other than of Weight, Time, Space factors only.

Flow substituting one of the other factors transforms the quality of movement

Incomplete effort and the different drives are very much a part of expressive movement, no matter whether this is deliberate or sub-conscious, as in shadow movements.

Bodily actions manifesting incomplete effort participation are expressive of a variety of inner attitudes. Of these we have learnt to distinguish six particular ones forming three pairs of opposites; these may be illustrated as follows.

Incomplete Effort		Emphasis on Movement	Giving Information about
1	(a)	Space and Time	Where and When
	(b)	Flow and Weight	How and What
2	(a)	Space and Flow	Where and How
	(b)	Weight and Time	What and When
3	(a)	Space and Weight	Where and What
	(b)	Time and Flow	When and How

It is difficult to attach names to these variations of incomplete effort as they are concerned with pure movement experience and expression. If the reader will take the trouble to perform some bodily actions with the two motion factors stated, but using one element of each only at a time, he might agree that the attitudes have the following characteristics:

1
   (a) awake { that is awareness, certain or uncertain, which may arise suddenly or gradually and may be embracing or concentrated;

   (b) dreamlike { that is unawareness which may be diffused or bold, and gloomy or exaulted;

2
   (a) remote { that is detachment which may include focus on self or universal attention, together with restraint or abandon;

   (b) near { that is presence which may have warm impact or careful consideration, or it may express strong attachment or superficial touch;

3
   (a) stable { that is steadfastness which may be resolute and stubborn or sensitively receptive. It may also be solid and powerful or delicately pinpointing;

   (b) mobile { that is adaptability which may be easy or sticky, slowly forthcoming or abruptly changing.

These attitudes appear very often as transitions between essential actions, and frequently have a recovery function. They play an important part in the configurations of motion factors and elements, which are groupings and not simply additions. These configurations build up individual units in which the single constituent part submerges entirely. Thus the whole gains each time a new meaning, importance and function which none of the single elements alone can claim to possess or to fulfil by itself.

Only groupings of effort elements can give information about their meaning

In considering the combination of three motion factors, we arrive at a basic set of new variations. These are usually observed when the expression is more intense, more pronounced or more communicative than in the display of inner attitudes.

Introducing "movement drives"

In a basic effort action the Flow remains latent and only the factors of Weight, Time, Space operate. When this is the case we speak of an "Action Drive".

Graph representing *Action Drive*

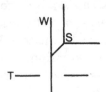

When Flow with either or both of its qualities (bound or free) replaces those of Weight, the drive becomes "vision-like", because it is now not supported by active weight effort and is therefore reduced in bodily import.

Graph representing *Vision Drive*

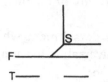

When Flow replaces Time, that means when there is no appreciable time quality, the expression becomes what we might call "spell-like". The inner attitude towards time rests and the movements radiate a quality of fascination.

Graph representing *Spell Drive*

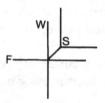

When Flow replaces Space and no particular attitude towards shape is displayed, that means when spatial qualities are dormant, bodily actions are particularly expressive of emotion and feeling. In this case we speak of a "Passion Drive".

Graph representing *Passion Drive*

In conclusion, it may be repeated that effort, with all its manifold shadings of which the human being is capable, is mirrored in the actions of the body. But bodily actions performed with imaginative awareness stimulate and enrich inner life. Therefore, mastery of movement is not only of value to the stage artist, but to everyone, since we are all concerned, whether consciously or subconsciously, with perception and expression. The person who has learnt to relate himself to Space, and has physical mastery of this, has Attention. The person who has mastery of his relation to the Weight factor of effort has Intention, and he has Decision when he is adjusted to Time.

Inner preparation of a bodily action

Attention, intention and decision are stages of the inner preparation of an outer bodily action. This comes about when through the Flow of movement, effort finds concrete expression in the body.

CHAPTER 4

The Significance of Movement

Travellers to countries inhabited by primitive tribes relate strange stories of rhythmic drum telegraphy. They tell how the news of an explorer's caravan passing near a village is transmitted in great detail at incredible speed to the remotest parts of the country by rhythmical signals given out by drum or tom-tom beats. One might be tempted to draw a parallel with Morse telegraphy where certain combinations of short and long vibrations signify letters that can be combined into words and phrases; but primitive tribes usually speak quite different languages and often have no knowledge of the sound-combination and rhythm of words of any tribe but their own. The rhythm must, therefore, have another significance which despite prolonged investigation has remained hidden from European minds. Rhythm seems to be a language apart, and the rhythmic language conveys meaning without words. Modern European races seem to be totally lacking in the intelligence capable of grasping the meaning underlying primitive rhythmic movements.

Primitive tribes seem to have a rhythmic language which conveys meaning without words

Investigators were grappling with this problem for many years until a native African gave a useful clue. The clue was that the reception of these drum or tom-tom rhythms is accompanied by a vision of the drummer's movement, and it is this movement, a kind of dance, which is visualised and understood. The method approximates to a science, and it is jealously guarded by secret societies. The sensitivity for movement observation amongst these primitive peoples has become a kind of native international language through which communication can spread across a continent, say from the east to the west coast of Africa, thousands of miles, with incredible speed.

This language could be compared with Esperanto and other constructions, if there were not the difference that these constructions are quite artificially invented, while the drum language, which is rather a kind of effort communication, has grown and developed through ages out of man's own rhythmic instinct. We have lost this language of the

Drum language, a kind of effort communication, a language of the body

body, and there is little probability that we can rediscover it, or, at least, make it practicable for use on the stage. Our approach to Nature and life differs all too fundamentally from that of our ancestors who first moved to the rhythm of the drum. Yet, there is a recrudescence in our time of the sense of rhythm and of the mime element in the theatre. It is beyond the aim of this book to give a detailed account of movement research, although what is said here is based on systematic investigation. A few aspects of the analytical and historical classification of movements have been included, however, as they may be of interest to the reader.

It has been found that bodily attitudes during movement are determined by two main action forms. One of these forms goes from the centre of the body outwards into space, while the other comes from the periphery of the kinesphere inwards towards the centre. The two actions underlying these movements are those of "gathering" and "scattering". Gathering can be seen in bringing something toward the centre of the body while scattering can be observed in pushing something away from the centre of the body. Gathering is a more flexible movement than scattering, the latter being more direct. The curve of the movement in a gathering action is preceded by an outward movement resembling that of scattering away from the centre of the body. This preparatory movement is, however, less stressed than the following inward gathering movement, which is the main purpose of the action.

Relationship between bodily carriage and gathering/scattering gestures of the limbs

Many combinations of these two actions can be performed by our limbs each working independently. For instance, it is possible that one arm gathers while the other scatters, and even within the arms it is possible that the upper part of the arm scatters while the forearm and hand make a gathering movement. In a scattering action the reverse is true; the pushing away from the body is the main purpose of the movement, and any preparatory approach near to the body is unstressed and of secondary importance.

In a step in which the weight of the body is transferred to one leg while the other is free, the free leg can either perform the next step or can perform a leg action which can consist either of a gathering or a scattering movement. It is possible to scatter with the heel, while the toes gather. As the upper arm and the hand can perform contrasting actions, so also can the legs and feet or any other two parts of the body. Certain combinations of gathering and scattering movements are characteristic of particular epochs of history and at certain places of the world.

Combinations of gathering and scattering in various parts of the body show cultural and period characteristics

The most natural way to move is the style of harmonious movement which was common in ancient Grecian and Roman cultures. Small diversions from this harmonious ideal never deform the movement to the extent that the attitude becomes disharmonious. The movement habits of primitive tribes, however, appear to the European mind and feeling as grotesque or disharmonious. The grotesque forms of Moorish dances were introduced at the time of the Crusades, but their asymmetric tortuousness was surpassed by movement fashions which can still be seen in the sculptures ornamenting our Gothic cathedrals. It was a mere accident that these masterpeices of architecture were ever associated with the barbarous Goths. In these picturings every limb of the body, every finger or toe, seems to be gathering and scattering simultaneously into different directions of space, producing its own action life. Again, here is a mentality poles apart from that of its predecessor, the Greco-Roman mentality, which expressed harmonious combinations of scattering and gathering action forms.

Without attempting to offer even a sketch of the history of movement, it can be said that in certain epochs, in definite parts of the world, in particular occupations, in cherished aesthetic creeds or in utilitarian skills, some attitudes of the body are preferred and more frequently used than others. Everybody can recognise not only the Greco-Roman ideal of movement which still moulded the court dances of Elizabethan times, but also the primitive attitudes of a more grotesque style which persisted in the popular jigs of the same epoch. It is easy to understand how the selection of, and preference for, certain bodily attitudes create style; yet, we must remember it is in the transitions between positions that an appropriate change of expression is made, thus creating a dynamically coherent movement style. We can connect bodily attitudes with either harmonious or grotesque transitional movements. The fact that any deviation from the main fashion or style of an epoch has been looked upon as abnormal and lacking in style, or that such deviations have ever been considered ugly and wrong, is due to a peculiarity of the human herd instinct. Communities seem to regard a certain uniformity of movement behaviour as indispensable for safeguarding the stability of the community spirit. They also tend to stress a common ideal of beauty, very often connected with a utilitarian value, especially esteemed in the community of a particular epoch. We see here how the attitudes of powerful warriors or hunters, of scholars or priests, artisans and draughtsmen, and also of the idle parasites of society at one period or another are considered to be the only beautiful and stylistically right attitudes.

Movement styles have been in some way useful in a particular period, or in a particular country, according to the main needs of the civilisation. Such partial aesthetical and utilitarian conceptions which lead people to say that this or that tennis-player, skater, batsman or film star "has style" descend frequently to tiny details of movement habits. The small difference of a gathering or scattering position of a foot, for instance, might provoke favourable or unfavourable opinion of the style of a champion athlete or a film star. This subconscious evaluation of people's

Historically seen: aspects which create movement style and need for one having dynamic coherence

The value of movement style to a community

Tiny details of movement habits create style

movement is practised by almost everybody. The artist, however, has to represent more than typical styles or typical beauty. He is interested in all the deviations and variations of movement. The representation of ugliness or clumsiness, or rather what is called so in certain periods of history amongst different peoples, belongs as much to the realm of the actor's art as the fashionable ideals of typical movements. Extraordinary movement combinations do very often fix the focal points of a dramatic conflict. The finer shades of style will be understood only after a thorough study of the rhythmic content of the attitudes in which a definite series of effort combinations has been used.

Evaluation of style requires study of the finer shades of effort rhythm

The tradition of ballet has preserved a great number of fundamental movement shapes which can be considered as symbolic actions. Such shapes are, for instance, the arabesques and attitudes. If we penetrate to the source of the significance of these shapes we find that it lies not in the final pose, but in the movements leading to it. With primitive dancers or with small children, it is the flexible flow of a gathering movement by which an object is gripped. This gesture of possession contrasts with that of repulsion, a direct scattering or pushing away movement. Possession and repulsion are fundamental urges. In traditional dancing these actions are petrified into characteristic body carriages or poses. The direct form is called "arabesque" and the flexible form "attitude". It is obvious that nothing has been left of the original gathering and scattering movement expressive of possession and repulsion, for in classical ballet both are sublimated into a gentle picture of a satisfied state of mind. The numerous arabesques and attitudes of ballet dancers have nevertheless an expressive kinship with the actions from which they originated, although they do not signify but symbolise similar inner strivings. In principle, a symbol does not mean anything definite, but it can call up a variety of images in the spectator. The fundamental urges of possession (grip) and repulsion (punch) which find expression in primitive movement are in ballet dissolved into a varied scale of harmonious forms that possess only an evocative character.

Gestures of possession and repulsion derive from basic fundamental urges

Classical dancer's "attitude" and "arabesque" have expressive kinship with "gathering" and "scattering"

The ordinary actions of everyday life, most clearly seen in working movements, constitute one stratum of the world of movement. A further stratum might be distinguished in the winks, nods and cries communicated during speech. Here conventional movements are substitutes for words. The dynamic arts of acting, singing and dancing represent a third stratum of effort expression. Movements performed in ballet have lost their connection with the primitive drives of man to such a degree that we relegate them to a realm akin to that of a dream state. In our dreams, movement is fantastical, though it may be linked to the forms of our everyday actions. Terror dreams, for example, show chaotically clustered effort expressions which may have some resemblance to the fundamental

Three strata of effort expression

Effort expression in dreams

urges of possession and repulsion. In harmonious dreams all fear of the struggle of life is dissolved into a smooth flow of effort as in elevation, floating or flying.

The differences between "attitudes" and "arabesques"

If we analyse the two sublimated forms of gripping and repulsing, we shall find that the poses called attitudes fill space much more completely than the arabesques. Attitudes show a relationship to all dimensions: high, deep, right, left, forwards and backwards. It is as if all space were raised into one comprehensive dimension giving a similar aesthetic impression to that of an orchid whose intertwining curves create a perfect unity. Attitudes are final poses which cannot well be further developed. Arabesques are of a quite different character. They do not fill space, but radiate from the centre of the body in definite directions, and are not so self-contained as the attitudes. Arabesques strive towards some point of the external world; they are not final. They demand a continuation of the motion in a clearly indicated direction. This penetrating of space which is characteristic of an arabesque contrasts with the confined centrality of the attitude.

The feel of movement is, in these two poses, concentrated on the symbolism which the shapes of space reveal to people sensitive to such impressions. In a state of concentrated inner attention almost everybody is sensitive to such impressions. Experience of the symbolic content and its significance must be left to the immediate comprehension of the

Verbal interpretation of a dance experience will remain unsatisfactory

person who watches the movement. Any verbal interpretation of this inner feeling will always be something like a translation of poetry into prose and will remain on the whole unsatisfying. But a concrete observation of movement based on realistic considerations of man's behaviour in time and space is possible and useful for the actor-dancer, and even for other people interested in the art of the theatre.

How movement material is used in pure dance and in theatrical dance forms

The art form in which visible movement expression is still cultivated in our time is stage dance. This includes on the one hand pure dance and on the other theatrical dance forms, such as ballet, mime and dance-drama. All these are real happenings so far as real actions of real bodies are concerned. Such movements show a more stylised form in dancing than in acting. In acting and miming, the movements have more resemblance to everyday behaviour. A savage or a child knowing nothing about the theatre might mistake the happenings in a play for happenings in real life. This would not occur in a dance where the movements of the dancers show, on the whole, such dreamlike transformations of everyday actions and behaviour that their symbolic character cannot be mistaken. Dance uses movement as a poetic language, whereas mime creates prose of movement. It is the arrangement of effort expression of everyday life into logical yet revealing sequences and rhythms which gives a theatrical performance its special character.

It is interesting to consider the aim of the transformation that *The silent world of* constitutes ballet. The world too deep for speech, the silent world of *symbolic action* symbolic action, most clearly revealed in ballet, is the answer to an inner need of man. If there were nothing in ourselves that could respond to this strange world, no one would ever want to witness ballet or dance. Interest in ballet becomes more comprehensible if we realise that the most deeply moving moments of our lives usually leave us speechless, and in such moments our body carriage may well be able to express what otherwise would be inexpressible. However, it is not the fashion today to regard mime and ballet as messengers of things ineffable, and people of the more sophisticated and cynical type are not likely to experience the kind of inner stirring which leaves them speechless. Ballet nowadays tends to be considered mainly as an exhibition of graceful motions and beautiful bodies. Yet the deeper interest in this world of silence cannot be said to be entirely dormant or dead. Many people, perhaps a greatly increasing number of people, feel that our waking lives are as full as our dreams of symbolic actions, and that some medium is needed in which actions of this kind can find aesthetic expression. That medium is obviously to be found in that area of movement which we call dance and mime.

But what are symbolic actions? They are certainly not just imitations or representations of the ordinary actions of everyday life. To perform movements as if chopping wood, or as if embracing or threatening someone, has little to do with the real symbolism of movement. Such imitations of everyday acts may be significant, but they are not symbolic. Man in those silent movements, pregnant with emotion, may perform strange movements which appear meaningless, or at any rate inexplicable. Yet, and this is the curious thing, he moves with the same actions which *Man uses the same actions* he uses in chopping, carrying, mending, assembling or doing any everyday *in working and in dancing* operation; but these actions appear in specific sequences having shapes *but rhythms and shapes* and rhythms of their own. Words expressing feelings, emotions, *in the latter have their* sentiments or certain mental and spiritual states will but touch the fringe *own life* of the inner responses which the shapes and rhythms of bodily actions are capable of evoking. Movement can say more, for all its shortness, than pages of verbal description.

Movements can, however, be named and described, and those who are able to read such descriptions and reproduce them might get the feel of the moods expressed by them. Single movements are, of course, only like the words or letters of a language; they do not give a definite impression or a coherent flow of ideas. The flow of ideas must be expressed in sentences. Sequences of movements are the sentences of speech, the real *Sequences of movement* carriers of the messages emerging from the world of silence. The question *are the carriers of messages* now arises whether any comprehensible order can be found in these *from the silent world* emanations of the silent world, and if so, whether this knowledge of orderly principles would be of advantage to the actor-dancer, and to the general standard of art of movement on the stage.

Theatrical art has its roots in the visible and audible performance of effort manifested through the bodily actions of the actor-dancer. Actor-dancers in performing all kinds of implemental and non-implemental movements are indeed real people engaged in real bodily actions; the silent world of ideas and inner stirrings lies brooding within these actions waiting to be formed into coherent shape. The performer stresses some of these strange shapes arising from the world of silence and stillness; and though he uses the same movements as any ordinary working person, he arranges them into rhythms and sequences which symbolise the ideas that inspire him. The average actor will admit only with reluctance that the enjoyment of his art by the public is based upon a subconscious analysis of his movements. Yet the fact is that the spectator derives his experiences from the artist's movements. In his own way, the viewer distils the material presented, although he does so to a great extent subconsciously. The form of the artist's presentation does suggest certain directions in which the spectator's distilling process can operate, but it does not totally determine this process. Even though only a few of the artist's movements have acquired a conventional meaning, it does not alter the fact that meaning is conveyed by movement. The exclusive use of movements with fixed meaning will never result in a work of art. It is just the unusual combination of movements which makes them interesting to the public.

The stage artist creates definite rhythms and shapes symbolising his ideas

Inspiration should be a less enigmatical word than it has been in the past, for it would seem that mankind is slowly learning to penetrate more deeply into the healthful world of silence. The ordinary, crude stagecraft becomes more and more intolerable. Shakespeare shows in several of his plays the contrast between acting resulting on the one hand from a penetration into the world of silence and on the other from the crude imitation of reality. It seems that we stand not only at the gateway of the discovery of a new form of acting, but that we must recover lost territory and regain the knowledge possessed by our ancestors centuries ago.

The job of the stage artist

The focus of the artist's interest in movement can be clearly discerned only if we try to assess what the job of acting or dancing really is. This job does not differ a great deal from any other human activity. A working person's job is to deal with material objects, and it is essential to know what to do with them. The working person might use his bare hands, or a set of tools. Hands are in fact nothing but tools attached to our bodies as living implements. The handling of objects and of tools involves movements which have practical results. The actor-dancer's distinction is that he works with no tool other than his body. Properties which he might use are not real tools or objects of his actions but are accessories to his movements. The use of the movements of his body includes, with the actor and singer, the use of the movements of his voice-producing organs. The artist of the stage has to exhibit movements that characterise a human personality's behaviour and growth in a variety of changing situations. He must know how personality and character are mirrored in gesture, voice and speech. He has, in using these movements, to communicate the ideas of a playwright or a dance-composer to the audience. He has to know how to get into contact with the spectator. He must help the team comprising himself and his fellow actors to establish the magnetic current between these two poles, stage and audience, and, as well as gaining a general mastery of his body and speech organs, he must also train and control his personal movement habits. The actor-dancer must be able to observe his own acts of movement as well as those of his fellow actors. The actor's motion-study is an artistic activity, and not a fact-finding device. In his study he will try to condense phases of effort into definite rhythms and shapes. The resemblance of his acts of movement to those of the personality represented can differ in exactitude as any other kind of portrait or description may differ. A portrait is never identical with its model. The model of the actor is frequently imaginary, but its characteristic traits can be extracted only from the observation of reality.

People's movement behaviour can be observed systematically, and it is useful for the artist of the stage to learn something about the best procedure in observation. The purposeful assessment of a person's *Movement observation* movement should begin at the moment one first sees him. A person's *in the service of* entrance through a door, his walk up to the first stop, and his carriage in *characterisation* standing, sitting, or any other position can be of great significance. It is the task of an artist in creating a fine and lucid characterisation not only to bring out typical movement habits, but also the latent capacities from which a definite development of personality can originate. The artist must realise that his own movement make-up is the ground on which he has to build. The control and development of his personal movement habits will provide him with the key to the mystery of the significance of movement.

Science tells us that motion is an essential of existence. The stars wandering across the sky are born and die. They wax and wane, some colliding with others, some burning themselves out. Everywhere is change. This ceaseless motion throughout measureless space and endless time has its parallel in the smaller motions of shorter duration that occur on our earth. Even inanimate things, crystals, rivers, clouds, islands, grow and dwindle, accumulate and break up, appear and disappear.

Motion becomes movement in living beings, who possess an inner urge *When motion becomes* to use time and the changes that occur in time for their own purposes. *movement* They develop into beings with individual traits, whether plants, animals or men. Man is ordained by reason of his material and spiritual needs to cultivate personal relationships with his fellows. In all these urges movement plays the paramount role. We see the movements of animals when they are building their abodes, gathering their food and slaying their enemies. We see religious dances in which human beings are praying to their gods. We also hear the movements of their tongues, throats, and chests producing the sounds when words are spoken and psalms and hymns are chanted.

Inner movements of thinking and feeling

The inner movements of feeling and thinking are mirrored in man's eyes and in the expression of his face and hands. The art of the stage developed from mime, which is the representation of inner movements by visible outer motions. Mime is the stem of the tree that has branched into dance and drama. Dance is accompanied by music, and drama by speech. Both music and speech are produced by movements which have become audible. Musical sound arouses emotion; spoken words express thought. But the musical quality of speech also colours words with emotion. For example, when one lover says to another, "Look at these fields!" the listener will know from the intonation that the words express an entirely different emotion from what peasants express, when, with minds on their work, they use exactly the same words: "Look at these fields." Further, speech is far more expressive when coloured by accompanying mime gestures. But in reality there exists no speech without bodily tension. Such tension is potential movement, revealing

Movement in mime, dance, drama, music and speech

sometimes more of a person's inner urges than do his words. Movement permeates miming, dancing, acting and singing. It is life as we know it. It is present also in the playing of instruments, in painting pictures, and in all artistic activity. In each case, movement is not only a physical fact, but a fact which varies in significance with its ever-changing expression.

The roots of mime are work and prayer. Work secures our material existence; prayer our spiritual growth and development. Work is akin to prayer when it is done not solely to maintain life but for a higher purpose. An artistic performance on the stage setting forth an ideal can thus be very near to prayer. Prayer can be closely allied with work. The building up of ideals, as is often done by fervent prayer, may be just as hard work, or even harder work, than manual labour. It may entail the full use of all our energy. The conflicts arising in work and prayer are represented in mime, drama and dance. In ancient times, dramatic poetry, and dance linked with music, developed from worship; and in modern times this is still the true order of development. At the dawn of philosophical thinking

Originally all art forms were a unity

all art forms were a unity; today they are separate disciplines. But even today drama is sometimes kin to dance and dance to drama; words can be filled with music and movement, while dance and music are frequently pregnant with the ideas conveyed by words. Vestiges of mime are found today in the puppet theatre, in clown scenes in the circus, and in pantomime, but outside these play-forms there is still sufficient scope for studying visible movement in drama and ballet.

It is true that the movements of an actor or singer are in these days seldom observed by the public. The fact that the actor or singer has to move from one place to another, that he or she is sometimes obliged to sit *Public's little awareness* down or rise up, or on occasion fall to the floor or into the arms of a *of movement in the* partner, or in other situations exhibit gestures expressive of his feelings, is *theatre* taken for granted. Provided the performer's movements and gestures are not clumsy, the public pays little conscious attention to them. Some people think that in theatrical performance movement is by far the least important; that decor, colour, light and costume matter a great deal more. Lack of style in scenery and dress is more severely criticised than want of style in movement. The pleasing bizarre colouring of the back-cloth obscures the unexpressive stiffness of the actor or singer, who nevertheless is bound to observe meticulous discretion in his bodily movements in order to be effective. It would be all to the good of some stage artists, and advantageous to large numbers of the public, were they to recognise that effective expression and restraint of movement is an art which can be mastered only by those who have learned how to give free rein to their movements.

The dialogue in drama, and the lyrics of song, can explain every *Discussion of the theatrical* happening, feeling and emotion; and the spoken word can be effectively *arts* communicated by radio. People enjoy broadcast drama, and few miss the objective trappings presented visibly on the stage or in the film. Without doubt, the listener to a broadcast misses least the visible movement. He is perfectly happy to listen to the audible movements of the performer's voice organs, and to hear the words. The audible movement of the vocal chords becomes almost a dance in singing. Like dance, which is a highly musical art, singing appeals more to the emotions than to the intellect. Ballet, of course, cannot dispense with the full deployment of visible bodily movement. Although music is an essential part of ballet, ballet music does not give the same pleasure when broadcast as when the ballet is seen on the stage. Good ballet music, which in itself satisfies the hearer, comes nearest to pure orchestral music, when the visual impression of movement is unnecessary. Orchestral and concert music in general does not properly belong to the theatrical arts, although music plays an essential part in opera. It is, however, interesting to note that orchestral music is produced by the most precise bodily movements of the musicians. It is not so important to see them, although many people derive great aesthetic satisfaction from the gesticulations of their favourite conductors and from the movements of the instrumentalists. In general, however, the musicians' movements are considered as working movements in which skill is more important than aesthetic performance.

As mentioned before, the theatrical art in which bodily movement is all important is mime, an art form which is almost unknown in our time. The nearest approach to it was the silent film before the "talkie" came. When the accompanying music intimately combines with the visible movement on stage or screen one can speak of dance-mime, which is something intrinsically different from dance. Mime uses a differentiated language of gesture, a language which can be fairly well translated into words. Mimed dialogues or soliloquies can be understood and verbally described, at least in their essentials. Such dialogues are rare in dance, in which the sinuous beauty of movement, with its accompanying music, are the main means of artistic expression. It is just as difficult to describe a dance in words as it is to interpret music verbally.

Dance-mime and dance compared

Attempts to find a common denominator for the significance of movement in spoken drama and in dance have caused many headaches. It is not so long ago that the fashion in acting suddenly changed from pompous gesticulation to a naturalism devoid of any movement expression at all. Playwrights, actors and directors became bored with the dancelike over-acting of an epoch saturated by melodramatic sentimentality and turned to the imitation of everyday life on the stage. But they were unable to appreciate the almost invisible finer movement tensions between people conversing in everyday life, and the immobility which they cultivated gave birth to a dead style of acting. These attempts to behave in a naturalistic manner resulted in a theatre devoid of movement. Since such passivity failed to touch the public, interest in the theatre declined. Audiences were disappointed because the thing they wanted to see was withheld. They wanted not only to hear but to see the expression of the various conflicts arising from human striving after material, emotional, and spiritual values; and such conflicts are most clearly expressed in movement.

Changing fashions in the style of acting and the role of movement expression

As neither the old untrue melodramatism nor the new boring naturalism could satisfy the justifiable expectation of the public, experimenting artists tried to get back to the sources of theatrical expression. One of the leading ideas was that an actor would never achieve the display of personality required in acting without having previously experienced full abandonment to the passionate human urge for movement as manifested in mime and dance. Inner tension, which is characteristic of the actor's art, was found helpful in that it imparted life to dead traditional ballet movement. In both cases, in the new art of acting, and in that of dancing, the fundamental rules of the old art of mime come again into the foreground. When mime, and with it the significance of movement, is entirely forgotten or neglected, the theatre is dead.

The preliminary researches and successes of a new attitude towards the mastery of movement on the stage would have remained problematical if a new factor had not widened the field of vision at the right time. This new factor is, surprisingly enough, movement as we observe it in industry. This fact becomes less amazing when we consider that all through history movement on the stage drew its inspiration from the occupational motions of the now most numerous part of the population, the workers. Generally speaking, there were early times when priestly rituals inspired the movement imagination of the actor and the dancer. Later on, the ceremonies of royal courts or the exercises of chivalry set the tune for dance and provided the plots and the movement style in drama. Now in our own time the industrial revolution has given new concepts of aesthetical beauty to all the arts, for the newly acquired knowledge of the workers' movements has led to a fresh mastery of movement on the stage. The procedures in training for skill and efficiency in industry show many parallels with the new training methods of the modern stage artist.

Industrial revolution gave a new inspiration to movement imagination

One factor is here paramount, and that is the discovery of the elements of movement. Whether the purpose of movement is work or art does not matter for the elements are invariably the same. The absolute congruity of man's working movements and his expressive movements is a staggering revelation.

Congruity between man's working movements and his expressive movements

The movements of a living being serve, in the first place, to secure the necessities of life. This activity can be summed up in one word: work. Not only man but all animals work in the search for food and shelter, in hunting, in contriving their abodes, and in caring for their own needs and the needs of their young. Work involves conflict. Living beings struggle with their surroundings, with material things, with other beings, and also with their own instincts, capacities and moods; but man has added to this the struggle for moral and spiritual values. The theatre is the forum wherein the striving within the world for human values is represented in art form.

In the art of mime the visible movements of the body are used as the sole means of expression; and it is obvious that working actions represented in mime are always the expression of ideas and have no real practical purpose. An actor might light a fire in order to convey the idea that the room in which he is playing is cold. He might be acting on a fine summer day when there is no need to light a fire; but he does it at the direction of the playwright in order to show that he is cold. His movements consist of bodily actions, taking pieces of wood, perhaps chopping them, and throwing them into the grate. He shivers, shakes his shoulders, strikes his body with his arms to speed up his blood circulation, but all this is make-believe. Actually he might be sweating, and would rather recline to slow down his throbbing heart-beat than accelerate it by exertion. The natural working movements in such a scene are an obvious imitation of the everyday actions, such as used in lighting a fire. Yet the surprising fact is

Working actions in mime express ideas

that these movements can also be used as expressive movements.* The movements made in chopping wood consist of repeated hits, using the forearm. Similar movements are made as the result of an inner excitement, of course, without hatchet and wood, to signify a desire to hit out at the cause of the excitement, as one would hit out at an adversary. The action of throwing wood in the grate can be exactly repeated without these accessories in a gesture expressive of the rejection of a disagreeable idea or proposal. A similar movement to that used in lighting a match might express a sudden and precise mental decision.

The basic actions of a working person are also fundamental movements of emotional and mental expression

Examples of the movements of everyday working actions that can be applied to the expression of inner states of mind and emotions are legion, but there exist a few such as pressing, thrusting, wringing, slashing, gliding, dabbing, flicking and floating which are the basic actions of a working person, and, at the same time, the fundamental movements of emotional and mental expression. Sounds and words are formed by movements of the speech organs, and these are rooted in the same basic actions. Sounds can be produced by pressing, thrusting or other movements of the speech organs, and will result in pressing and thrusting accents of the words. Such accents are expressive of the inner mood of the speaker.

As an experiment, the reader might try to pronounce the word "no" to express different shades of meaning, and he will then easily recognise that he can say "no" with the following actions, each producing a different sound quality and expression:

> "No" with Pressing Action — firm, sustained, direct.
> "No" with Flicking — gentle, sudden, flexible.
> "No" with Wringing — firm, sustained, flexible.
> "No" with Dabbing — gentle sudden, direct.
> "No" with Thrusting — firm, sudden, direct.
> "No" with Floating — gentle, sustained, flexible.
> "No" with Slashing — firm, sudden, flexible.
> "No" with Gliding — gentle, sustained, direct.

Connection of visible and audible movement

By accompanying each of these sound expressions with a gesture of the quality indicated, the reader will become aware of the connection between audible and visible movements.

The discovery of elements of movement identical in work and expression sheds a new light on the significance of visible and audible movement on the stage. The actor and the dancer will certainly have some advantage

* The reader is advised to *perform* the movements described in this and the following chapters. It is only by bodily experience (exercises in which actions are repeatedly tried out) that mastery of movement can be achieved. It is not necessary to use objects or properties since these can be imagined. The main stress is always on bodily action.

in knowing more about the movements that will help them to picture characters in conflict. But just as a painter is not yet an artist if he knows no more than the composition of his palette, so the actor or dancer will need more than a mere outline of basic actions and their classifications. The actor or dancer, using movement as a means of expression, will rely more on the feel of the movement than on a conscious analysis of it.

Knowledge about the elements of movement an advantage

The observation and assimilation of a person's characteristic movements is analogous to the recalling of a musical theme. If a person can remember a tune, and is capable of reproducing it mentally, he will later be able to discern single pitches and rhythms in every detail. Without falsifying the proportions of the strength, duration, and other peculiarities of the flow of the music, he will be able, with his inherited or acquired musical memory, to write down, to discuss and to analyse the tune. The ability to observe and comprehend movement is a like gift, but it is also, as in music, a skill that can be acquired and developed through exercise.

It is essential that the student of stage movement should cultivate the faculty of observation, which is much easier to do than is generally believed. Actors, dancers and dance teachers usually have this faculty as a natural gift, but it can be refined to such a degree that it becomes of the utmost value for the purposes of artistic performance. It is obvious that the artist's procedure in observing and analysing movement, and then in applying his knowledge, differs in several respects from that of the scientist. But a synthesis of scientific and artistic movement observation is highly desirable, for otherwise the movement research of the artist is likely to become as specialised in one direction as that of the scientist in the other. Only when the scientist learns from the artist how to acquire the necessary sensitivity to the significance of movement, and when the artist learns from the scientist how to bring order into his own visionary awareness of its inner meaning, can a balanced whole be created.

Movement observation should be practised but a synthesis of a scientific and an artistic approach is desirable

CHAPTER 5

Roots of Mime

Life as lived and as mirrored on the stage consists of a chain of happenings. Some links in the chain, specific situations or actions, are more important than others. Dramatists and choreographers select the situations and acts which they judge to be most necessary to their themes, and focus upon them the spotlight of the stage. This means that, in terms of speech and bodily actions, movement qualities with their rhythmic, dynamic and spatial structure are of special importance as through these the performance becomes articulate. This is the creative part of the performer's work, which must, however, be adapted to the content of the play, if not anyway following a set script of composition.

Through rhythmic, dynamic and spatial structure movement performance becomes articulate

The public needs this articulation of thought, feeling and event. It is essentially what they want to see and experience, since the course of life does not always allow us to contemplate the origin and consequences of all our acts. We are therefore indebted to the dramatist and to the actors who mirror these happenings for us. We go to the theatre and the cinema in order to see human life in leisurely contemplation and, as it were, through a magnifying glass. This magnifying activity is, however, not a simple intensification of expressive movement. Exaggerations of intensity can lead to overacting and this should naturally be avoided. The precision and clarity of the form of movement are more important than the intensity of the performance.

Movements spring from man's striving after material and spiritual values

Man shows in his movements and actions the desire to achieve particular aims and ends. We may refer to these as values which could be of a material or spiritual nature. In the theatre, the spectator, consciously or unconsciously, looks for guidance about values for which man strives in various situations. To see such strivings so represented on the stage is an education for all who are not entirely insensitive to the conditions of life and human fate. If this is true of the spectator, it will surely be so also of the performer. He must know how to mirror life conditions and their fateful outcome in selected bodily attitudes and movement qualities; otherwise, the values which he wants to convey remain unrecognisable. The performer must learn to read and exhibit the behaviour of the people he meets. He must also be able to imagine them in different situations and must acquire the art of expressing human striving after values in movement. The plots of plays and choreographies may excite the actor's imagination by offering situations and conflicts, but he himself has to select and stress those movements which place essential values in the right perspective. All this has little to do with psychology as generally understood. The study of human striving reaches beyond psychological analysis. Performance in movement is a synthesis, i.e. a unifying process, culminating in the understanding of personality caught up in the ever-changing flow of life.

Movement performance is a unifying process

Some people who bypass their fellow men, ignoring their struggles, sufferings and joys, miss a great deal of the meaning of life and what it offers. They miss the opportunity to experience what is hidden below the surface of existence, and they tend to ignore the theatre where the depths are revealed. They lack the sense of the significance of persons and situations, and the world appears to them, more often than not, an accumulation of meaningless happenings. It is not for such people that the actor-dancer performs, and he must be careful not to fall into the same indifference. A person who has no interest in his fellow men's strivings is not an actor, hardly is he a human being. The emptiness of life resulting from such a lack of interest and sympathy spells blindness towards the most important values of human existence. Blind people are largely shut out from the life that surrounds them and much of its significance. These unfortunates are unable fully to appreciate the great variety in the flow of life and its values. Though they themselves are immersed in it, the part they play in life is limited to their enfeebled capacity. But it must be observed that blind people often develop other, and most astonishing, faculties that serve them and the community well. People of inner vision are aware of the flow of life, and the creative contribution they can make to it. They are eager to witness in the theatre the mirroring of life's happenings which enriches their understanding of situations, characters, and relationships.

No human action is without consequence. An act once done perpetuates itself in an infinite chain of happenings that would never have occurred but for its committal. "Things are what they are and the consequences will be what they will be", as the great English writer, Butler, observed. Man creates his fate more or less consciously, but the acts and omissions of his fellow men interfere with and modify the particular individual's creative struggle. The clash of personalities in given situations weaves an extremely complicated pattern, the outline of which continually changes. Conflicting passions and winning tendernesses, harsh partialities and anxious hesitations, create a maze of relationships which cannot be completely disentangled or understood by the analysing intellect in isolation. It is the privilege of theatrical art to help the spectator to comprehend the acts of life in their completeness, and to awaken in him the capacity to relate this maze of action to his subconscious desire for values. People strive for something that has value for them. This value can be material, or moral, or spiritual, but it remains a value. The desire for values precipitates conflicts, either in the inner life of the individual, or in the external world between persons who cherish different and incompatible values.

The dramatist and the actor, as well as the musician and the dancer, confront the spectator with situations and actions in which the greatness or littleness of the struggle after values, and the quality of the values striven for, evokes partisanship. The spectator adopts an inner attitude towards what he sees and hears. In extreme cases he is moved to love or hatred (or he may remain indifferent), not of the actor or the character represented, but of the whole complex of values involved in the dramatic sequence. Even when amused by comical situations, these decisive inner attitudes towards values strongly emerge. The motions of bodies and sounds seen and heard on the stage stir the imagination, and awaken the will to look with open eyes into that vaguely discernible world, the world of human values. This is the reason for the theatre: the desire to make acquaintance with the world of values. The question which the dramatist and the actor have to answer, however, is not "what is valuable in the course of life?" but "how do the strivings for values which form the movement patterns of action intertwine and coalesce into individual and collective fates?"

By its nature the theatre is a mirror, and not an institution for moral judgment. The value of the theatre consists not in proclaiming rules for human behaviour, but in its ability to awaken, through this mirroring of life, personal responsibility and freedom of action.

Our present form of civilisation has perhaps greater need than any earlier one to be awakened to the appreciation of values. The speed of modern life is not only little adapted to quiet contemplation but the feeling for values seems to be steadily atrophying. Many of the old institutions that endeavoured to enshrine moral values in civil and ecclesiastical law are decried, and the individual is left more and more to himself to make his own inner decisions, and to work out the development of his personal responsibility.

Theatrical art uses movements of the body and of the voice organs to mirror the trend towards values and the conflicts arising therefrom. The urge for values is revealed by the artist on the stage by means of expressive movement, that is movement of the body and movement of the mind.

What has to be made clear through movement is:

(a) the characters of the persons represented;
(b) the sort of values for which they strive; |
(c) the situations developing out of the striving.

With reference to the first point, it might be said that characterisation can be rough yet typical of masses of persons, or refined and related to an individual. The greed and furtiveness of, say, a thief could be typically represented by grasping hands with claw-like fingers exhibiting in arms and body cramped strength in a sustained and twisted manner. This could be contrasted by light and free movements of the victim, the honest man, for whom a direct and leisurely ease of body carriage may be more typical.

Through such and, of course, much more detailed characteristic effort stresses, something of an individual's life history can be shown. The circumstances and the inner disposition through which a man or a woman has become a thief are infinitely varied. The actor-dancer will have to lift out the particular ones of the character he portrays and formulate his effort sequences and interplay accordingly. In terms of movement thinking, the greed of the thief which we characterised above roughly by tight-clawing and twisted movements can evolve from any pattern of effort combinations. It can, for instance, develop from a fundamental attitude in which fighting against Weight and Space is prominent, suggesting a kind of stolidness and stubbornness as a general disposition. Through the particular rhythm of recurrences and variations of, as well as releases from, the basic effort character, the personality of the thief is further defined and his possible development indicated. The actor-dancer's movement imagination might find many different combinations and sequences of effort expression to characterise the specific nature of different dishonest individuals.

The development of a character can be in the portrayal through interaction of appropriate effort sequences

A more complex characterisation of an honest man may contain effort features which are apt to facilitate and even encourage the scheming of a thief. A frequent recurrence in the honest man's efforts of an onward-going flow in a leisurely and straightforward fashion might easily provoke the thief to develop opposite characteristics which underline his crooked-ness. On the other hand, in reaction to dishonesty the honest man might conceivably be roused to anger, changing his straightforwardness into the firm twist of a wringing, gnarling action.

A person's effort display may influence that of another one

The second point, concerning the kind of values for which a person strives, compels us to classify values. This may be done first in a general way. The striving for material values or goods can roughly be contrasted with the striving after mental or spiritual values. Continuing our example of dishonesty and honesty we can say that material goods have a value for both the honest and dishonest man. Both have respect for property,

Different kinds of values

but to the honest man the greater value might easily lie in the idea of property rather than in its acquisition. The thief, however, who covets another person's property and subsequently steals it does so because he values it as a material gain. The idea of property is a legal conception, and as such a mental or intellectual possession. It becomes of spiritual worth when the defender of the lawful right to property desires to protect the despoiled person from the despoiler, and when the main motive is justice and commiseration.

Here we have three kinds of value after which people may strive and each will require a different method of movement presentation. The basic set of effort combinations used will in each case show a different accent and selection of body areas as the main carriers of the expression — i.e. hands, face, trunk and legs in characteristic gesture, carriage and step. There will also be a different manner of unfolding all these in space, both in relation to self, i.e. concerning the shaping of gestures and bodily carriage, and in relation to the immediate surroundings, i.e. concerning patterns of locomotion.

A person's leanings are revealed by the kinds of effort combination, spatial unfolding and body areas selected

The defender of his own rights and possessions might engage in actual bodily fight with the thief. In his desire to recover the stolen object his effort display might very much resemble that of the thief. But were he to defend the rights of another person and be fighting from a feeling of commiseration and indignation, he might, first of all, not engage in an actual physical fight at all. The vehemence of his movements is more likely to flow onwards freely in his hitting, punching kinds of gestures rather than lead to a deliberate clenching, twisting characteristic of acquisitive action. The main difference will, however, appear in the expression of satisfaction when the purloiner is overpowered. The pleasure derived from regaining a lost possession will again differ from that felt in the performance of an act of compassion. The action drive which is likely to be prominent in the hand-to-hand fight of the defender of his property with the thief will certainly be transformed into one or another of the other effort drives. This means that the motion factor Flow replaces one of the other factors, lifting the movement out of its matter-of-fact physical workaday expression into one which displays more feeling and inner qualities.

Example of action drive transformed into one or another movement drive

Now let us look at the third point, that of a situation in which the struggle for values takes place. We shall perhaps do this best by pursuing the former example we have chosen out of the infinite multitude of events in life, namely, that of an honest man struggling with a thief, for it contained the striving after both material and spiritual values. Here is then a scene presenting situations which might develop through the struggle of the various characters for different kinds of value.

A poor old woman is trudging along a street. She buys a hot pie from an itinerant vendor. On leaving the barrow she drops her purse but is unaware of her loss. A man following her picks up the purse realising that it is hers, but he puts it in his pocket thinking no one has seen him do it. He joins his companions at the street corner. The old woman discovering her loss returns in panic and searches frantically for her purse. She approaches the group of men at the street corner and asks if anyone has seen it. They only jeer at her. Moving away in despair, the old woman is stopped by a pedlar who advances towards the group of men and singles out the man whom he rightly judges to be the thief. The latter is uneasy at first but gains courage from the support of his cronies. Meanwhile the old woman has gone off, but during the ensuing argument she returns accompanied by a policeman. The men at the corner flee. The pedlar holds the thief who throws down the purse before the old woman and is thereupon arrested. The poor woman weeps with joy at the recovery of her solitary treasure.

Example of situations arising from striving after different values

In this scene the passage of time created by a series of situations allows, for instance, an actor to show the pedlar's capacity of a feeling of commiseration. This will have to be expressed in effort sequences before, during and after his encounter with the thief. At first, significant effort accents might occur in shadow movements accompanying the pedlar's selling actions. These can be developed into characteristic incomplete effort transitions between his actions of stopping the old woman, approaching the group of men and seizing the thief. His satisfaction that the poor woman received her purse back might again be expressed in his shadow movements, this time not accompanying any purposeful action, but significantly qualifying the posture of a man who has succeeded in upholding inner values of pity and compassion.

Suggestion how shadow movements and incomplete efforts may qualify action

It may be observed that there can be many variations of the themes of theft, defence, fight, triumph and every one requires careful formulation of the movement expression according to character, value and situation. There is great merit in inventing such scenes and in trying to experience and exhibit in movement the subtle differences of behaviour that well up from man's innermost nature and which form the basic content of mime.*

We now offer some material for the invention of mimes concerned with strivings after other inner values. The conscious observation and experience of effort arising in the body will be of great importance for the growth and refinement of the capacity to mime, to act and to dance.

Both a kind person and a strong brute could subjugate anyone

---

* The reader should try to perform and differentiate the streetseller's behaviour;
(a) as described in the above scenario; and
(b) as it may have been, had the purse been stolen from the street vendor himself.

physically weak or otherwise delicate. But a kind person respects delicacy and despises brutality, while the brute enjoys power and its material advantages and delights in overcoming the powerless. It is obvious that the brute will have a body carriage quite different from that of the amiable person. Yet there is a difference in the carriage of the brute who merely desires to overpower his victim and the brute who delights in inflicting mental, moral, or spiritual torture. Although there is no question of a material object, if the victim's body is not considered a material object, there is nevertheless a difference in the struggle for values, and this will express itself in the use of varying kinds of effort. It is the difference between a plain conflict and subtle torture. This is brought out in the following scene.

A dejected parlourmaid is tidying a room and preparing the table for her master. The master enters and sits down at the table completely ignoring her. The maid brings in his breakfast, and in her nervousness drops a dish spilling some of its contents on her master's coat. He is furious and strikes the girl, causing her to cower under the blow. In rushing from the room, the maid meets her mistress, who perceives the cause of the trouble but for the moment remains silent. The mistress picks up the broken fragments and sits down. A short time after her husband goes out; the wife calls in the maid. She enters terrified, but is calmed by her mistress. It is obvious that the master and mistress will differ in their effort display not only during the scene sketched in this example, but habitually in their everyday life movements.

The characters in a drama are ordinary people exhibiting movement combinations in a specific form. The collision between characters, and the display of the various tragic and comic solutions of their conflicts, which is the stuff of dramatic art, become visible in changing movement behaviour; and these changes can sometimes tell a story without dialogue, or verbal explanation. Almost any scene can be performed dramatically with dialogue, or as a mime-scene without dialogue. Dance, in which the movements are often co-ordinated with the rhythm of music, is, of course, different because dance does not require dramatic content. Yet there is also a good deal of characteristic change of movement behaviour in non-dramatic dance. Historical dances, folk and national dances that have no dramatic content are frequently used in drama to stress the period style of the play. This is often taken as a basis for ordered mastery of movement in acting. Dances are not just inserted in the plot or story as a special form of entertainment; the movement of the dance colours the action life of the play and determines its movement style. For example, Shakespearean presentation was largely influenced by the movements of the Elizabethan court-dances of French and Spanish origin, and by the then contemporary popular jig which permeated the spirit of almost all

Effort display in physical conflict and mental torture

Movement behaviour in dramatic art and in dance

comic scenes. Yet we find in Shakespeare's plays a broad range of natural action and passion which seems to demand for its portrayal, in some places at least, the full scope of human effort configurations uninfluenced by the specific movement restrictions of the period.

Great dramatists and composers of ballet never devise their works purely to mirror an epoch. They depict man in his ageless struggle: the conflict of his natural and personal passions restricted by the passing movement fashion of the period. The dramatist needs and uses both the period style and those forms of unrestricted effort in which a person's real character breaks through. Every individual has the tendency to enlarge the range of his effort capacities. This enlargement is connected with his own personal development. The public, and therefore the dramatist, is very much interested in the development of personality, and this depends on an increase in the range of effort capacity. *Development of personality depends on increase in range of effort capacity*

The observation of effort in the service of characterisation penetrates man's inner attitudes from a different angle from that used in the investigation of period styles. When studying a person, the artist will soon notice that the observed movements vary in significance and importance. The full record of the whole movement behaviour will be beyond the capacity of the observer unless he is highly skilled in systematical assessment. It is, however, possible to learn to observe some of the most important developments in the movement behaviour of a person and to note those of immediate interest. It can be seen, for instance, that the observed person has a peculiar manner of using various parts of his body. With a sitting person it will be mainly the upper part of the body, trunk, arms, hands and head, together with the various parts of the face that will call for attention. The motions of the voice-producing organs will be mainly visible in the mouth and lips; they will be audible in the pitch and rhythm of the voice. Special attention should be paid to the movements of the eyes, eyelids, and eyebrows. Glancing in various directions, the movements of the eyes can be accompanied by the turning, lifting, or lowering of the face. In other cases the head may remain immobile, the eyes alone moving. *Some indications how to observe movement*

The movements of the fingers as they manipulate or toy with objects have to be distinguished from those of the hand. It is advantageous to disregard the objective aims of the movements, and to confine attention to the hands or other parts of the body irrespective of whether the action performed is or is not practical. The character of the stirrings is best expressed in terms of movement, that is in Space, Weight, Time and Flow elements as revealed in bodily actions. These are the key to the understanding of what could be called the alphabet of the language of movement: it is possible to observe and analyse movement in terms of this language. The investigation and analysis of this language of movement and *Alphabet of the language of movement is basis for analysis* therefore acting and dancing can only be based on the knowledge and practice of the elements of movement, their combinations and sequences, as well as on a study of their significance. The shapes and rhythms which are formed by basic effort actions, movement sensations, incomplete effort, movement drives, give information about a person's relation to his inner and outer world. His mental attitude and inner participation are reflected in his deliberate bodily actions as well as in the accompanying shadow movements.

Movement language conveys a person's relation to his inner and outer world

It can be observed that all practical actions are preceded by four phases of mental effort which become visible in small expressive bodily movements. There is the phase of *attention* in which the object of the action and the situation of its execution are inspected and considered. This may be done with direct concentration or, more fleetingly or perhaps with circumspection, in a flexible manner. This, in the normal sequence of mental effort, is followed by a phase of *intention* which may range between strong and slight. The kind of muscular tensions produced in small body areas will give information about a person's determination to act. The intention to make an active movement might be abandoned before it is carried out. For example, a person's attention is drawn to a book lying on a table. He stands and looks at it directly. His intention of reading it becomes visible in a certain muscular tension in his chest and neck. He decides to pick up the book and his hand moves swifty towards it, but before he does so he remembers that he has something else to do. No longer having the intention to read the book, he drops his arm.

Practical actions preceded by four phases of mental effort: attention, intention, decision, precision

In this example we have already introduced the next phase, namely that of *decision*, which in this case was shown by a sudden jerk in the hand. A decision can, of course, also gradually be arrived at which would become visible in a more sustained, slow stir in some small part of the body. Before the objective action starts there is yet another phase which can be observed. This might tentatively be referred to as that of *precision*. It is that very brief moment of anticipation of performing the actual deed which, very often, if unfamiliar, is highly controlled by a bound flow effort, or, if the opposite is the case, is unconstrained and charged with free flow.

These four phases constitute the subjective preparation of the objective operation and are mostly closely condensed and may then be transferred, either partly or fully, into the action carrying out the job. It is, however, possible that they occur simultaneously or that their sequence is reversed, varied or complicated, or that even one or the other phase is omitted.

Shadow movements inform about inner processes

Shadow movements tell us about such inner processes; and many of a person's most characteristic movements are those which he does unconsciously and which precede, accompany or shadow his deliberate actions. A person might scratch himself, rub his chin, pinch his nose, shake or jerk his fingers, or perform other vague gestures having no definite practical significance but which are done for movement's sake. We might refer to these as subjective gestures. There are other kinds of movements, for example, nodding, pointing, winking, waving, which replace words such as "Yes", "There", "Here", "Look out", etc. We call them conventional gestures. All these gestures usually display incomplete effort and it is only under excessive excitement that they will appear as a full-scale thrust, glide or other basic action. *Subjective and conventional gestures*

Conventional gestures, unconscious shadow movements, expressive and functional actions deliberately done are mixed in a person's behaviour in a most complex manner. They might appear in any kind of sequence, and sometimes several kinds of movements or actions might occur simultaneously.

Understanding of movement comes through discovering which attitudes towards the motion factors prevail or are absent in a sequence of movements. The same element might be present in almost all actions, while others may occur only in some actions and some may be missing altogether. It is as if one should say "this picture is mainly blue and that picture mainly red." But there can be in the blue picture spots of other colours, perhaps of red, though further colours, say, yellow or green, might be lacking. This means that although the movement expression of a person can be governed by one sort of action, for example, gliding, it is still possible for other actions to be present, particularly those which are immediately akin to gliding, such as pressing, floating and dabbing. Opposite qualities such as slashing or hitting can be seen especially in compensatory movements which are made unconsciously. *Prevalence or absence of movement elements show a person's effort make-up*

The prevailing action character and its shadings can be observed either in the reactions of a person to specific situations, or in his or her habitual movement behaviour. For some people certain reactions will be quite improbable, or at least exceptional, but rarely entirely impossible. We can hardly expect a relatively weak or slow person to be heroic or quick except in exceptional circumstances. It is here that we reach a crucial point, namely the possibility of change in habitual effort make-up. Slow changes can be promoted through a conscious understanding of the structure and rhythm of one's habitual effort patterns, but some of them may be so ingrained that it is very difficult to modify, to extend and thus to change them. *Can habitual effort patterns be changed?*

When depicting a character, an actor has not only to mirror the general effort make-up of the character, but he must also be able to convey the development of that character's inner attitudes during the happenings of the play. Some characters might remain unchanged, presaging their tragic or comic fate. Others will develop either in a positive or negative direction, and their adaptation to these new situations might be the *The art of characterisation* essential feature of the play. In this respect the change of habitual effort make-up expressed in bodily carriages and actions will be one of the essential means by which the actor builds up his characterisation. Characterisation is an art, and, apart from psychological truth, there exist many factors which determine the artist's choice of definite sequences of movement and effort. In his first intuitive approach to his role the actor as well as the dancer might be unaware of the particular sequences of effort qualities he chooses. Yet in the process of formulating his part, giving it shape and remembering it, he has consciously to select movement phrases, rhythms and patterns.

In addition to portraying a single person's character, it must be remembered that the clash or harmony between characters is in reality the clash or harmony between effort patterns.

Such dramatic situations are created by a kind of effort chemistry which often results, as in chemistry proper, in a destructive explosion or the creation of new compounds. The correlation of effort patterns of different persons *vis-à-vis* one another is a special aspect of the study of mime, which uses a kind of effort dialogue similar to the spoken dialogue in drama. An effort dialogue between, for example, a gentle person and a brute has to reveal details of both their characters, especially of their attitudes towards certain values. The kind person might be charitably moved to help and protect the powerless. For him charitableness may be of higher value than his natural gentleness. He could appear to be brutal when confronting his protégé's oppressor. A brute, however, might, in addition to being cruel, find great satisfaction in his cruelty. Cruelty is for him of higher value than just overpowering his victim; and in his cruelty he might assume deceiving suavity.

Disguising and revealing efforts

In the effort dialogue between the two contrasted types, the brute and the kind man, many shades of effort combinations will arise. The brute becomes soft and seemingly gentle to achieve his vicious propensities; the kind man becomes hard and seemingly brutal to achieve his charitable aim. Such disguising efforts are frequently the prelude of the revealing effort, because a brute's softness must finally change to his habitual hardness, and the kind man's hardness must finally change into his habitual gentleness. Change from an indulging attitude to a fighting one involves an increase in force or speed or straightness in direction, and that from a fighting to an indulging attitude an increase in the movement sensations of lightness, duration or expansion. The change may take place in one factor after the other or in two or three factors at the same time.

Probabilities of change in the chemistry of human effort

There is much more probability of abrupt change when the brute, from an occasional softness, falls back into his habitual harshness than when the gentle person tries to transform his natural tender manner into a robust one. The sudden anger of a habitually jovial and gentle person will dissolve slowly; the slowly developing softness of a habitually brutal person will show sudden recurrence to his true character. Such probabilities are not rules, because the conflict within the various effort configurations is so very complex.

The fact is that man is able to disguise his effort nature as well as his effort patterns to a certain extent. Such a disguise will deceive only an unobservant person, though leaving him with a vague feeling that something is wrong. This feeling results from subconscious observation, or rather from observation not brought to full consciousness. Any effort of another person which is perceived mainly by our eyes or ears, causes an effort reaction that can, but need not always, result in an easily perceptible visible or audible movement. The counter-effort can be twofold: it can be either similar or dissimilar to the effort causing the reaction. A brute could meet inner hardness in a friendly person whom he harshly attacks, but this hardness might slowly dissolve into softness as the kind man begins to pity the benighted mind of his attacker. The kind person's softness might find a vague sympathetic response in a brute, but this is quickly transformed into an inner and external harshness.

In drama, many nuances of effort qualities appear in transitional movements. They often show an incongruous interplay of rhythms and shapes indicative of conflict between a character's inner attitude and his outer demeanour. As stated before many times, inner attitudes are manifested in the movements of small areas of the body and are often barely visible. The well-known expressiveness of the eyes is occasioned by contractions and relaxations of the muscles of the eyes. These contain complex series of basic and incomplete efforts, affecting the pupils and the movements of the eye-balls. As soon as the impulses of the nerves and muscles within the eye-sockets spread to the eyelids, the eyebrows, or other face muscles, the inner effort becomes externally more visible and controllable. The effort can spread to the larger muscles, and finally to all the muscles of the body, in which case it is visible as body carriages, gestures and steps. The eye muscle reactions are, however, not the only ones which express inner reactions. Breathing can be influenced by startling impressions, and even the heart-beat can be so affected. Glandular changes may add to the effects of effort response. All these inner impulses may take visible form in small or in large muscle reactions.

Inner reactions mostly visible in movements of small body areas, e.g. eyes

Physiological effect of effort response

All effort action or reaction is an approach towards values, the primary value being the maintenance or achievement of the balance needed for the individual's survival. The maintenance of this balance demands the reciprocal functioning of body and mind according to a person's effort capacities or character. Balance requires adaptation to situations and a more or less clear recognition of the values towards which the individual habitually or only occasionally strives. To discriminate between material

Effort is primarily spent towards maintaining the individual's survival

values and mental values is not always easy. The kind person in our example above is believed to reach after mental values; the brute to strive mainly after material goods. Comforts, sensuous pleasure or even happiness are material values, but they could be the foundation of mental values which might demand the cessation of acts that could occasion pleasure and happiness, if such acts should involve moral obliquity.

Estimation of material and mental values

The brute who renounces what others would consider comfort and pleasure to glut his lust for cruelty does not seek material advantage only, not at least as material advantage is commonly understood. He might indeed lose his life in the attempt to assuage his passionate cruelty. He might exhibit enormous courage and consider it a value exceeding the satisfaction of his urge to cruelty. Such vacillations in the estimation of material and mental values during the development of a situation must make the artist careful in the use of moral and ethical precepts. The artist has not to judge; it is his business to portray; and from the point of view of stage representation he must concentrate on the expressiveness of his bodily actions. The main point is to realise that any aspect of the struggle for values, no matter how low or how high they may be as judged by the generally accepted scale of values, can be expressed in the language of movement. It is not just the simple bodily action of relinquishing things to others, or of depriving others of them, that is expressed by the multiple interplay of effort patterns, but also the motives behind this behaviour.

Effort observation; motives and consequences of our acts

Through effort observation we do not only witness past happenings and motives, but we glimpse into the future, into the probable consequences of our acts. We can detect certain fundamental incongruities in effort configurations which if extended will lead to disaster. This can be avoided through the development of harmonious effort patterns. Such balance can be brought about by awakening in the individual a feeling for appropriate transitions, rhythms and forms within movement sequences.

The shapes and rhythms of our movements are powers by which practical actions can be performed, but they also contain strong generating energies which give rise to reactions that carry beneficial or disastrous consequences. For example, charity can become egoism, if the pleasure of giving in friendship, love or sacrifice exceeds the pleasure of making other people happy. Inner impulses wishing to disguise egoism become visible in shadow moves. The warmth of a gesture may be contradicted by the cold stare of the eyes, or the twitching movements of face muscles. One part of our body may assent, another part deny. We may breathe heavily or excitedly while otherwise displaying an external calm. The struggle of effort impulses within ourselves is part of the drama. Almost all our decisions are the result of an inner struggle which can become visible even in an entirely motionless body carriage. Bodily position is always the result of previous movements or the foreboding of

Contradictory effort patterns

Stillness of bodily carriage and posture is a result of inner impulses

future movements that leave or foreshadow their imprint on the body carriage.

A mime often transmits to the spectator what kind of an inner struggle his character is going through, solely by his body carriage or posture without perceptible movement or sound. Even in everyday life it is possible to see in a person's carriage as well as in his movements the way his thinking and feeling goes. It is the mime's task to draw us, his *Task of the mime artist* audience, into the world of his drama by his bodily expression and gestures so that we can identify ourselves with the characters and suffer with the suffering, or feel angry with the rightfully indignant, or laugh at our mirrored selves. If we can do this, we have been lifted out of ourselves and removed from the selfish pleasure for the sake of which we often help people or give presents or perform other charitable acts. Although the artist draws for his creation on real-life situations, feelings and actions, these are not directly depicted in his mime but are given significant form out of his own imagination and vision. In this sense the actor can be a great giver of his own self, and become a mediator between the solitary self of the spectator and the world of values.

The mediating activity of the actor demands veracity in a high degree. The competent actor, mime or dancer strikingly reveals the possibility of expressing the values of veracity and all their complications through bodily action. It is a great error to regard the theatre and acting as make-believe, and as dealing with false actions and ideals. Mime and the theatre introduce the spectator to the realities of the inner life and the unseen world of values. The attempts to understand the world through naturalism and materialistic realism are doomed to failure. The realities of the inner life can only be depicted by art in which reason and emotion are compounded, and not by intellect or feeling in isolation. To give the right *Reason and emotion* answer to the innermost expectations of the spectator, the actor must *compounded in art;* master the chemistry of human effort, and he must realise the intimate *expectations of the* relationship between that chemistry and the struggle for values of which *spectator* life consists. Although a spectator might not have any other reason for visiting the theatre than the wish to be entertained, he is nevertheless dissatisfied if he does not glimpse the realities of the world of values, and that world can be effectively depicted only through both external and inner mobility.

It has been stated that the change from indulging to fighting involves quickening of the actions performed, or giving them increased directness or strength. It has also been stated that the reverse change from fighting to indulging involves a growth of duration, expansion and lightness, resulting in increasingly sustained, flexible and gentle actions. This can be tried out by imagining or performing scenes in which such hardening or softening occurs. The fighting against or indulging attitude towards a

The two opposite attitudes towards motion factors, fighting and indulging, are basic in the emotions of hatred and love

motion factor form the basic aspects of the psychological attitudes of hatred and love. So it is useful if the artist realises how these two poles of emotion are related to other forms of inner attitude, and how their relationship is mirrored in the movements of different characters. The emotion of love might be symbolised by the image of a goddess whose fundamental movement behaviour shows toleration and acceptance of the motion factors of Weight, Time and Space. The qualities of her movements, will, therefore, mainly derive from the basic effort action of floating and show a predominance of gentle, sustained and flexible elements. But hatred, with its hard impact upon reality, could be symbolised by the image of a thrusting demon who, in his typical acts of movement, fights against all the three motion factors of Space, Time, and Weight. His motions will be direct, sudden and firm. Such mythological characters can, of course, be represented on the stage. It will not be too difficult for an imaginative actor-dancer to depict them. Even should he be a modern cynic, he will remember the age-old symbolism of love's soft floating movements, and of the violent and abrupt movements of hatred.

Admixture of plain and more intricate effort habits explained

The characterisation of a mere mortal will be more difficult, because imagination credits gods, goddesses, and demons with plain and uncomplicated effort habits, whereas those of mortals are seen to be much more intricate. There are types of people, for example, the go-getting, careerist politician, who might either be a lover or a hater. Sometimes he will pose as the one and sometimes the other, just as his personal interest dictates. But he will mainly show in his movements an effort admixture, deriving from that of the demon, when he defends the cause of an oppressed group of people, which may be less from a feeling of compassion for them than from the desire to overcome his hated political opponents. Or take for instance a more charitable person, "the good samaritan", who fights rather to succour the oppressed than to overcome the oppressor. His movements derive from an entirely different inner attitude than those of the politician, namely that of smoothing out the effects of strife and discord. His effort admixture will be nearer that of the goddess.

Kinship and crosscurrents in effort patterns

The reader will understand the kinship of the politician with the thrusting demon, and of the good samaritan with the floating goddess. Why then does not every individual politician look and behave like a predatory animal, and every charitable person like a lamb? Simply because so many crosscurrents are in the effort patterns that characterise personality. Using the terms of movement thinking, an attempt can be made to define the mixed effort characteristics of our politician and our samaritan, at least in an elementary way.

Starting with an action in which fighting against certain motion factors prevails, it can be said that a *presser*, a *slasher* or a *dabber* fights against two motion factors, and indulges in one.* They have two points of hatred and one point of love to their credit. The meticulous dabber, the rigid presser and the flighty slasher have also some love in them which prevents the easy change into full-blown monsters of hatred. A politician, a good samaritan, indeed anyone, might have any of these various mixtures of effort qualities. Here we shall attribute them to the figure of the politician and show three varieties of him. *Fighting two motion factors, indulging in one*

In some effort actions, one can observe indulgence towards two motion factors together with a fighting attitude to the third. People addicted to such effort combinations are the *gliders*, the *wringers* and the *flickers*. These are the blemished lambs in whom, for example, love scores two points and hatred one, and the harmony of the character is disturbed by the fight against one motion factor. One can imagine a smoothly gliding good samaritan rather over-anxious to succour, or one desperately wringing his hands uncertain what to do, or a third charitably flicking the dust from the unfortunate victim and fluttering excitedly about him. Always there will be something lacking in these gliders, wringers or flickers that keeps them on a fallible human level so that they never reach the rather static ideal of perfection of a goddess or a demon. *Fighting one motion factor, indulging in two*

The above examples of effort analysis are designedly crude simplifications, but they may be of help to students in showing them how to sum up the effort components of an individual character and how to use this in the service of characterisation. In schematic presentation the characteristics mentioned may be registered as follows.

Who	Fights against			Indulges in aspects of			Action Character
	Weight	Space	Time	Weight	Space	Time	
Demon	firm	direct	sudden	—	—	—	Thruster
Goddess	—	—	—	gentle	flexible	sustained	Floater
Politician A	firm	direct	—	—	—	sustained	Presser
Politician B	firm	—	sudden	—	flexible	—	Slasher
Politician C	—	direct	sudden	gentle	—	—	Dabber
Samaritan A	—	—	sudden	gentle	flexible	—	Flicker
Samaritan B	—	direct	—	gentle	—	sustained	Glider
Samaritan C	firm	—	—	—	flexible	sustained	Wringer

The scheme is brought nearer to life if one understands that:

(a) no individual persists always in one and the same effort quality, but changes continuously;

*See Table below.

(b) during the changes some effort elements are
    (i) kept intact, while
    (ii) some become over-stressed, thus altering the quality of the effort more or less visible, and others get almost entirely lost.

In the changes indicated under (a) any sequence might arise according to circumstances. Many situations can be imagined in which, say, Politician A changes his original mood of pressing into that of a softer one, say, into one characteristic of Samaritan B. Anybody can start with any of the basic action moods, no matter whether it is habitual to him or not. He can then with greater or less effort mobility run through whatever scale of moods he likes, or which outer circumstances compel him to assume. Each set of such changes is an effort phrase which speaks a clear language conveying rises and falls, hesitations and precipitations; and these, if frequently repeated, bring out habitual characteristics.

Changes of effort quality form effort phrases

This chemistry of effort follows certain rules, because the transitions from one effort quality to another are either easy or difficult. In ordinary circumstances, no sane person will ever jump from one quality to its complete contrast because of the great mental and nervous strain involved in so radical a change. If a person changes briskly from a pressing mood (Politician A) into a flicking mood (Samaritan A), it is a sign of great inner tension or excitement. Such situations can of course arise, but they are always, as it were, danger signals of inner high tension. Sometimes, the acting person reverts to incomplete elemental actions, in which only one or two elements are charged with inner participation, while the rest of the effort manifestation remains mechanical and is not supported by any inner attitude. A person showing this undecided form of effort expression could be, for instance, a *presser-glider* and be considered a rather vague character. Such a quality in a gesture might indicate anything, but its essence is the renouncing of strong action. The curious thing is, however, that characters are met with who, either habitually or in a special mood, use incomplete effort. Such moods of inactivity, arising occasionally, point to a depressed disposition of mind that inhibit the rise of full action drive. If these moods are constant, a character will be doomed to lifeless inactivity.

Transitions from one effort quality to another follow certain rules

The kinds of change are indicative of mental disposition

One can easily imagine the enormous range of effort combinations placed at the artist's disposal. The possibilities of changing and varying movement expression are innumerable. It does not simplify matters that all these numberless effort combinations can appear in movements of the limbs or of the whole body. But the basic idea is nevertheless simple enough. If one observes people's effort expression, it will be noticed that it is always one exactly describable movement containing one well-definable effort content that happens at one time. It is, of course, a different thing to choose one of the many possibilities of movement and effort content in order to characterise a person's behaviour in a specific situation on the stage.

Characteristics of movement behaviour can be found in one definite movement with definite effort content

The actor in observing people might recognise certain types of habitual effort, such as the *dabber-glider*, or the *slasher-wringer-floater*, which might offer a good basis to build up a picture of an imaginary person's movement behaviour. But how will the person behave in a particular situation? The effort make-up of a person observed at a particular occasion can give a sufficient picture to recognise some of his prevailing capacities. It can be reasonably expected that certain combinations of effort qualities might work out towards particular behaviour in definite circumstances. It is probable that a cautious and self-centred person in whose effort patterns no special forcefulness has been observed will, if confronted by a sudden violent attack, be inclined to retire than to stand his ground. The attacker might then modify his *thrusting* mood in order to reassure the cautious person.

Method of mutating basic actions

Punching or thrusting is relatively quick. If one modifies (that means slows up) the speed of the movement, thrusting is transmuted into pressing. When the strength of the thrusting is diminished, thrusting becomes a gentle dabbing. Thrusting occurs always on a straight line; but gradually curving it into an increasingly wavy line thrusting is finally transmuted into slashing. Such mutations can happen in an appropriate situation.

Each one of the basic actions can, through change in its speed, or its degree of strength, or the curvature of its path, be modified more and more until it finally becomes one of the other basic actions. This change can be compared with the grading one into another of the colours in a rainbow. As the many shades of colour can be understood as transitions or mixtures of the basic colours of the spectrum, so also can the great variety of actions observed in our movements be considered and explained as transitions or mixtures of basic actions.

Let us here investigate the effort range which is within the natural potentials of the previously mentioned Demon whom we gave the fundamental characteristics of a "thruster". Then we shall set against it that of the Goddess whom we characterised as a "floater", and compare the two. In this investigation it is, of course, assumed that the characters are entirely harmonious within their own type.

At first it may, however, be useful to survey the various factors of movement and their implications as set forth at the end of Chapter 3. This will be done in a tabulatory form as the obviously needed detailed explanations, particularly about the significance of movement and its complex ramifications in human expression, cannot be dealt with here.

Earlier in this chapter (page 104) we looked into the four phases of

*The significance of motion
factors in mental effort*

mental effort preceding purposive actions, namely those of attention, intention, decision and precision. In our tabulation we attempt to attach each of these to one of the motion factors and consider them not only as preceding an action but also as accompanying it.

The motion factor of Space can be associated with man's faculty of participation with *attention*. The predominant tendency here is to orientate oneself and find a relationship to the matter of interest either in an immediate, direct way or in a circumspective, flexible one.

The motion factor of Weight can be associated with man's faculty of participation with *intention*. The desire to do a certain thing may take hold of one sometimes powerfully and firmly, sometimes gently and slightly.

The motion factor of Time can be associated with man's faculty of participation with *decision*. Decisions can be made either unexpectedly and suddenly by letting one thing go and replacing it with another one at given moment, or they may be developed gradually by sustaining some of the previous conditions over a period of time.

The motion factor of Flow can be associated with man's faculty of participation with *precision* or, perhaps otherwise expressed, progression. This is his ability of attuning himself to the process of accomplishing, i.e. relating himself to the action. He may control and bind the natural flux of this process or he may give it an unrestricted and free run.

In Table VII shown opposite a survey is given of the Motion Factors and their combinations indicating how man's total self is involved when moving.

If we refer now to a character as being basically a "thruster", what does this imply from the movement point of view? What is at the bottom of the mutations of this basic effort action, and from what kinds of inner participation, attitude, or drive do they spring? The reader is encouraged to investigate the intricacies of psychological functioning on the basis of the movement considerations given in Tables VIII and IX.*

*Through effort analysis
possible meeting points of
opposite characters are
revealed*

A similar analysis of the natural potentials of the "Goddess", to whom we ascribed the basic effort action of "floating" as the characteristic quality, will show that in every respect she represents the opposite of the "Demon" (*see* Tables VIII and IX). Yet the mutations of which each should be capable will show through what kind of movement adaptations they can establish meeting points with their opposite.

*See pages 116-19.

Table VII

Survey of the Type and Meaning of Motion Factors and their Combinations

A. *One Motion Factor*

Motion Factor:	Space	Weight	Time	Flow
Inner Participation:	Attention	Intention	Decision	Progression
Concerned with:	Where	What	When	How
Affecting man's power of:	"Thinking"	"Sensing"	"Intuiting"	"Feeling"

B. *Two Motion Factors*

Motion Factors:	Space Time	Weight Flow	Space Flow	Weight Time	Space Weight	Time Flow
Inner Attitudes:	awake	dreamlike	remote	near	stable	mobile

C. *Three Motion Factors*

Motion Factors:	Space, Time, Weight	Flow, Time, Weight	Space, Flow, Weight	Space, Time Flow
Movement Drives:	Action (Flow dormant)	Passion (Space dormant)	Spell (Time dormant)	Vision (Weight dormant)

A careful study of the tables might give the reader guidance on how to develop movement sequences not only in producing a particular character in a play but also in creating poetic expression of various moods in dance.

Of course, characters and moods do not always develop in harmonious patterns. More often than not, especially in dramatic situations, unrelated qualities spring up side by side, creating disturbance and tension. It is, therefore, important to learn to master not only gradual mutations of a basic action, but to practise contrasting ones without easy transitions or balancing factors. In connection with this let us remember that effort changes are not always created by situations. The contrary is also true. As previously indicated, new situations are often created by effort changes of individuals and crowds.

The chain of happenings which is the very stuff of dramatic action, and therefore also of mime, has its roots in the chemistry of human effort. The nourishing soil of the tree of mime is the world of values.

THE DEMON

TABLE VIII

Basic Effort Action and its 3 mutations	Belonging to the Inner Drive of	Containing variations on Inner Attitudes	CHARACTERISTIC INNER PARTICIPATION WITH			
			Attention	Intention	Decision	Progression
A (basic effort action) Space : direct Weight : firm Time : sudden	Action	stable	direct	firm		
		near	direct	firm	sudden	
thrust		awake	direct		sudden	
1st mutation Space element is changed i.e. Direction is diffused through influence of sensation "pliant"		stable	flexible	firm		
		near		firm	sudden	
slash		awake	flexible		sudden	
2nd mutation Weight element is changed i.e. Resistance is weakened through influence of sensation "light"		stable	direct	gentle		
		near		gentle	sudden	
dab		awake *as in A*	direct		sudden	
3rd mutation Time element is changed i.e. Speed is slowed down through influence of sensation "long"		stable *as in A*	direct	firm	sustained	
		near		firm	sustained	
press		awake	direct			

3 transformations of original Basic Effort Action and their mutations

	Belonging to the Inner Drive of	Containing variations of Inner Attitudes	CHARACTERISTIC INNER PARTICIPATION WITH			
			Attention	Intention	Decision	Progression
Transformation I	Passion	(i) dreamlike		firm		bound or free
		(ii) mobile			sudden	bound or free
Mutation (a) Weight element is changed	{ Here Flow has re-placed Time element which remains dormant }	dreamlike		gentle		bound or free
Mutation (b) Time element is changed		mobile			sustained	bound or free
Transformation II	Vision	(i) remote	direct			bound or free
		(ii) mobile *as in I (ii)*			sudden	bound or free
Mutation (a) Time element is changed	{ Here Flow has re-placed Weight which remains dormant }	mobile *as in I (b)*			sustained	bound or free
Mutation (b) Space element is changed		remote	flexible			bound or free
Transformation III	Spell	(i) remote *as in II (i)*	direct	firm		bound or free
		(ii) dreamlike *as in I (i)*				bound or free
Mutation (a) Space element is changed	{ Here Flow has re-placed Space element which remains dormant }	remote *as in II (b)*	flexible			bound or free
Mutation (b) Weight element is changed		dreamlike *as in I (a)*		gentle		bound or free

TABLE IX

THE GODDESS

Basic Effort Action and its 3 mutations	Belonging to the Inner Drive of	Containing variations on Inner Attitudes	CHARACTERISTIC INNER PARTICIPATION WITH			
			Attention	Intention	Decision	Progression
A (basic effort action) — Float — Space : flexible, Weight : gentle, Time : sustained	Action	stable	flexible	gentle		
		near		gentle	sustained	
		awake	flexible		sustained	
1st mutation — glide — Direction is increased		stable	direct	gentle		
		near		gentle	sustained	
		awake	direct		sustained	
2nd mutation — wring — Resistance is increased		stable	flexible	firm		
		near		firm	sustained	
		awake *(as in A)*	flexible		sustained	
3rd mutation — flick — Speed is increased		stable *(as in A)*	flexible	gentle		
		near		gentle	sudden	
		awake	flexible		sudden	

CHARACTERISTIC INNER PARTICIPATION WITH

3 transformations of original Basic Effort Action and their mutations

Belonging to the Inner Drive of	Containing variations of Inner Attitudes	Attention	Intention	Decision	Progression
Passion	(i) dreamlike		gentle		bound or free
	(i) mobile			sustained	bound or free
	(ii) dreamlike		firm		bound or free
	(ii) mobile			sudden	bound or free
Vision	(i) remote	flexible			bound or free
	(i) mobile *as in I (ii)*			sustained	bound or free
	(ii) mobile *as in I (b)*			sudden	bound or free
	(ii) remote	direct			bound or free
Spell	(i) remote *as in II (i)*	flexible			bound or free
	(i) dreamlike *as in I (i)*		gentle		bound or free
	(ii) remote *as in II (b)*	direct			bound or free
	(ii) dreamlike *as in I (a)*		firm		bound or free

Transformation I

Mutation (a) Weight element is changed

Mutation (b) Time element is changed

{ Here Flow has replaced Space which remains dormant }

Transformation II

Mutation (a) Time element is changed

Mutation (b) Space element is changed

{ Here Flow has replaced Weight which remains dormant }

Transformation III

Mutation (a) Space element is changed

Mutation (b) Weight element is changed

{ Here Flow has replaced Time which remains dormant }

CHAPTER 6

The Study of Movement Expression

The best way to acquire and develop the capacity of using movement as a means of expression on the stage is to perform simple movement scenes. First the student should become fully conscious of the character of the person to be represented, the kind of values after which he or she strives, and the circumstances in which the striving occurs. Then, as a part of his creative function as a performing artist, he must select the movements appropriate to the character, the values and the particular situation. This selection involves intensive work. Improvisation of the acted scene, however brilliant, is not enough, nor is it sufficient to memorise a seemingly effective movement combination. What is necessary is that the student should, so to speak, get under the skin of the character to be portrayed, should penetrate the various possibilities of rendering the scene, and should analyse everything in terms of movement. He may find that some movement sequences will be easier to perform than others. If movements are chosen which are felt to be difficult, the student should try to find out the cause of the difficulty before attempting to master them.

Performing artist's creative function

There are two fundamental causes that obstruct an easy mastery of movement: physical and mental inhibitions. In Chapters 2 and 3, advice was given on the best way to overcome a general physical inability to move. The physical inhibition in a definite movement will best be removed by a sufficient number of repetitions of the bodily actions found difficult to perform. Great care should be taken to notice the effort qualities contained in the action, and to perform them clearly and exactly with the appropriate rhythms. The performance of effort sequences requires a related inner concentration and attitude. The causes of failure to perform certain combinations are, therefore, of a mental as well as of a physical kind.

How to achieve mastery of movement

The elements of movement when arranged in sequences constitute rhythms. One can discern *space-rhythms*, *time-rhythms* and *weight-*

Three kinds of rhythm

120

rhythms. In reality these three forms of rhythm are always united, though one can occupy the foreground of an action.

Space-rhythm is created by the related use of directions resulting in spatial forms and shapes. Two aspects are herein relevant:

(a) the one in which there is successive development of changing directions; and

(b) the other where shapes are produced through simultaneous actions of different parts of the body.

The first could be compared with melody and the second with harmony in music. Each requires a different flow of effort participation, for certain effort sequences lead to held positions while others develop movement lines of ever-changing angles and curves.

Space-rhythms lead to spatial forms and shapes requiring effort participation

In his daily work man learns to appreciate that he can handle certain things in one position better than in another. He will then turn his body until the object is in the best position for his effort. His experience allows him to map out the space around his body.* With the help of this imaginary map, he finds his way through all the effort combinations necessitated by his work and general behaviour, and in the end will find himself as much at home in this realm of effort as in his native town of which he knows every street, possibly every house and many of its inhabitants. The study of movement expression involves a mastery and understanding of space-rhythms, a brief survey of which is given in the second chapter. This survey is by no means complete, but it will stimulate anyone who tries to think in terms of movement to invent variations embracing more and more of the almost endless combinations of space rhythms that are possible. There is, of course, an inherent order, but it is beyond the aim of this book to discuss this.

Besides the space-rhythm of movement, we must consider its *rhythm in time*. Man's attitude towards time is characterised on the one hand by fighting against time in sudden, quick movements, and on the other hand by an indulgence in time with sustained, slow movements. Rhythms produced by bodily movements are characterised by a division of the continuous flux of movement into parts, each of which has a definite duration in time. These parts of a rhythm can be of equal or unequal length. In the latter case, some parts of the movement are relatively quick while others are relatively slow.

Continuous flux of time is made rhythmic through actions of longer and shorter durations

The significance of the time-rhythms of movement can be observed in individual dancers who have clearly discernible preferences for special rhythms. While one dancer will be more tempted to interpret music in

* See *Modern Educational Dance* by R. Laban, Macdonald & Evans, third edition 1975.

which the sharp metricality of regular beats prevails, another might be repelled by the exact metricality and prefer the free, irregular unfolding of time-rhythm. The precision of the metrical dancer is in strong contrast to the expressiveness of the dance-mime-actor preferring free rhythm. There exist many shades between the two extreme contrasts of regular and irregular rhythm. To a certain degree it is true that the dancer's legs and feet prefer metrical function; but feet, arms and hands should be equally able to express the qualities of a free time-rhythm. In fact, the whole body should be able to express the regular and irregular vibrations and waves of movement. Although an understanding and appreciation of music, which is an abstract expression of movement, can help the actor in his grasp of rhythm, it is not in itself sufficient. Even the dancer, who interprets music, has to translate it into the effort sequences from which his or her expressive steps and gestures arise. In dance the rhythm of movement is mainly expressed by the steps and this is particularly true of traditional ballet which uses a number of basic steps and characteristic combinations of these.

Free and regular time-rhythms should be mastered in expressive gestures and steps

Enormous erudition has been expended in constructing and reconstructing from old pictures and documents the exact shapes and rhythms of the dance movements and steps of past epochs. The oldest rhythms of which we have knowledge are those of ancient Greece and these, in the main, are related to poetic and dramatic works. From a purely historical point of view it is interesting to realise that the notation and interpretation of effort rhythms had already been attempted thousands of years ago. The Greeks have attributed to rhythms a definite significance, mostly expressed by an emotional mood. The combination of one short time unit with one long seemed to them to give the impression of masculine energy, while its contrast, a long followed by a short, was considered as the expression of femininity. A long and two short were felt as grave and serious, while two short and one long expressed the moderate tempo of a simple march. Excitement evoking terrifying states of mind was expressed in one long, one short followed by one long time unit, and two long and two short time units suggested violent agitation or its contrast, profound depression. Drunkenness, languor, and despair were suggested by a rhythm consisting of two short and two long. In associating short and long duration with the *weight-rhythm* of accented and unaccented parts of a movement sequence, six fundamental rhythms have been found. The chart on the opposite page shows these six fundamental rhythms, and tells the special significance of each.

Ancient Greeks attributed a definite significance to rhythms

Weight rhythms are combinations of stressed and unstressed parts of a sequence

The Greeks considered all other rhythms to be variants of one of these six fundamentals. These rhythms, called measures, were arranged in verses, strophes and poems. They considered rhythm to be the active principle of

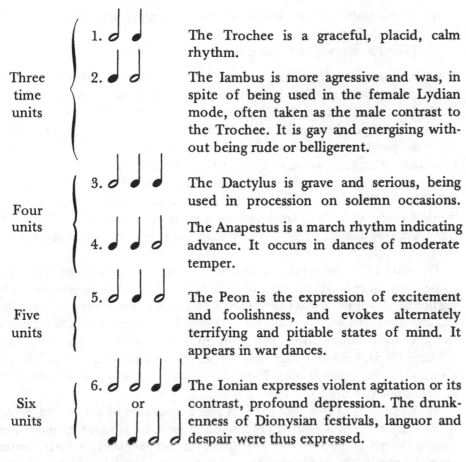

Three
time
units

1. The Trochee is a graceful, placid, calm rhythm.

2. The Iambus is more agressive and was, in spite of being used in the female Lydian mode, often taken as the male contrast to the Trochee. It is gay and energising without being rude or belligerent.

Four
units

3. The Dactylus is grave and serious, being used in procession on solemn occasions.

4. The Anapestus is a march rhythm indicating advance. It occurs in dances of moderate temper.

Five
units

5. The Peon is the expression of excitement and foolishness, and evokes alternately terrifying and pitiable states of mind. It appears in war dances.

Six
units

6. or The Ionian expresses violent agitation or its contrast, profound depression. The drunkenness of Dionysian festivals, languor and despair were thus expressed.

vitality. In music, they took rhythm as the male and melody as the female principle. Combinations of these rhythms had special associations in the Greek mind, as for instance,

a composition of the Trochee with a following Iambus signifying a mixture of moods as represented by these two rhythms. The Greeks knew many moods: austere, belligerent, festive, voluptuous, tender, passionate, enthusiastic, and supernatural. These moods were expressed by mixtures of the fundamental rhythms as follows.

(a) Austere, belligerent, rude movements of masculine character.
(The Dorian mood, using Dactylus, Anapestus, and Peon.)

(b) Voluptuous, flowing, tender attitude of feminine character.
(The Lydian mood, using the Trochee, Iambus and Anapestus.)

(c) Enthusiastic, religiously enhanced, passionate attitudes of a super-natural character.

(The Phrygian mood, using the Ionian and Peon.)

What is of special interest for us is that these moods were associated in the Greek mind with a definite flow of the time-weight elements of effort.

Habits of moods and alleviation of difficulties arising from them

Modern industrial workers' actions are very often confined to one or other of the fundamental rhythms determined by the ancient Greeks. They do not just express moods, but create habits of moods if frequently repeated. Watching workmen departing in the late afternoon from factories, one can recognise the rhythms which they have exercised during the day in the flow of their tired or excited shadow moves. Modern motion study and effort training have a great opportunity to alleviate the difficulties and tedium of work arising from the non-understanding of rhythmic capacity.

The student of movement may find one sort of rhythm more difficult to perform than another. He might discover that he has a preference for particular rhythms or rhythmic habits. Rhythmic deficiency of any kind can be corrected by training, by accustoming oneself through repetitive exercise to produce those forms of rhythm which at first are found

A dramatic scene contains, beside rhythm of movement, a greater rhythm of values and situations

difficult. It is, however, not only the rhythm of movements which is of interest to the actor-dancer. There exists a greater rhythm of values and of situations in a dramatic sense.

In the following five study scenes some examples of simple indications of characters, values and situations are given. The student of movement should try to analyse and to perform these scenes until he feels content with the result. The student must realise that there exist no right or wrong forms of an interpretation. It is for the artistic temperament and taste to find out which interpretation is preferred. Just as the painter risks the pleasure or displeasure of the critic viewing his picture, so the actor has to risk the acceptance or non-acceptance of his interpretation of a scene. The control of a skilled producer can act as a guide for the development of the students' taste, but it cannot be a substitute for personal discrimination.

Imagine and perform the following scenes.

Scene of Emotional Character

You have been asked to perform a scene of an emotional character such as: falling in love; quarrelling; observing a humorous situation; being faced with pain, suffering, or loneliness; expressing enthusiasm; getting drunk, sleepy or anything else which seems to you personally difficult to depict. Very well, get up and perform the scenes one after another, and repeat them until you feel the movements come easily; but beware of selecting all too easy movements to express these emotions. Analyse the effort sequences you use in each scene. Your anger, your pain, your gaiety, your enthusiasm and your sensual abandonment can be expressed in very different rhythms. Try to find out how different characters would show these emotions, and how the rhythmic movement would change with the different characters.

Scenes of Unusual Surroundings

You are now asked to imagine for yourself situations in unusual surroundings, such as: imprisonment; being lost in a dark forest; a man-of-the-street in a palace, or a rich man relegated to the hut of a beggar, or a prudent person in a doubtfully conducted night club, as well as other situations of your own invention. Proceed exactly as with the earlier specified emotional scenes: perform; exercise; analyse; relate them to definite characters.

Note: Characters can be: emotional; occupational types of different social classes; people of different age-groups, not omitting small children; people physically or mentally ill; people in various temporary stages of extravagant behaviour.

Scenes Containing Practical Actions

You are invited to perform scenes (without props) denoting activity, such as: household work, gardening, thieving as a pickpocket or a kleptomaniac, operating surgeon, nurse, clerk, film operator, hunter, or warrior.* The procedure is again the same: perform; repeat several times; analyse; and relate to definite characters. You will find that these four procedures will overlap to some extent. In performing you may analyse and attribute movements to a character and you will try to repeat parts of the performance. It is not wrong to do so, but afterwards the student should concentrate on a deeper and more detailed analysis. He should then study more thoroughly the suitability of his movements to the characters. Try to perform some of the above scenes in a tragic, comic or a burlesque manner.

Mime-dialogues

A good way to control and refine your performance is to suppose that a director is watching you. Imagine yourself in the chair of a critical director and try to see your performance through his watchful eye. Then return to your place and mime all over again the corrected version of the scene as you think it should be. It is also possible to perform a mime-dialogue by putting yourself first in the place of one and then of another

* Make all the movements clearly discernible for a spectator.

person acting with you. Deal with such dialogues exactly as with the other scenes: repeat, analyse and change your own character in an appropriate way. Invent short scenes of mime-dialogue in which different characters meet in various situations.

Scenes Concerning Movement and Costume

It is today usual to teach the actor-dancer certain forms of body carriage or dance steps supposed to have been appropriate to the movement fashion of a particular historical period. Some of these forms, especially those of the last few centuries of European history, have been transmitted from generation to generation. Our small amount of reliable knowledge concerning the form of the original movements is mostly derived from pictures representing scenes from the contemporary life of the period, and from the costumes of the time. Some of the dances of former epochs have been recorded in ancient forms of movement notation. Period costumes are a guide in concluding what the actual dance movements looked like; but the essential feature of the movement expression, the effort rhythms, can only be guessed. Yet it is advisable for the student of movement to become acquainted with illustrations of the costumes of different historical periods. By this means he is enabled to imagine the wearing of them and to adapt his movements to them.

A valuable exercise is to enter a room and to walk in a fashion characteristic of a period. Satisfaction or displeasure with the costume can be expressed in movement. The student attempting to wear costume in an appropriate way should comprise various styles: Savage styles, Antique styles, Medieval, Victorian, Contemporary styles as worn by peasants and townspeople, poor and rich, in everyday or festive attire.

Other Sources of Movement Stylisation

It seems that for several hundred years two kinds of opposing movement styles can broadly be discerned, that of the upper and that of the lower classes with perhaps a style intermediate between these two. To this day we can see this division in the cinema. The fashionable male and

Natural and fashionable movement behaviour and their effort admixtures

female film stars indulge in a movement convention which differs greatly from the average movement behaviour of the man in the street. We do not know whether this was also the case in the theatrical arts of remoter epochs, but it is probable that what is called period style is mostly the fashion of the well-to-do minority of a community in a specific historical era, while the natural behaviour of the majority has changed very little throughout the whole history of mankind. This natural behaviour is determined at all times by a rich mixture of the fundamental effort actions; they form the substance of man's movements no matter whether they are functional as in the service of work or whether they are expressive of mental and emotional states. It would be an interesting task to study the behaviour of men and women in period dances, and in the observation of religious rites, or at social gatherings. An attempt should be made to discover from the imagined wearing of period costume the movement forms used at those periods in all community assemblies, including, of course, dance assemblies.

A few remarks about the significance of bodily carriage thus become necessary, as in changing outer attitudes we can detect fundamental characteristics. A body stretched upwards will give a different impression from one curved downwards. The mood of the person moving in these two contrasting directions, no matter whether stretched or curved, can be the same. In hilarious mood, accompanied by laughter, the whole body, and both arms, might be stretched high in the air, or the hilarious one could curve and crouch, clasping his abdomen with both hands. Again the mood of anger can be expressed in either a high or a low body carriage, yet both these attitudes have a different significance. Such typical contrasts in body carriage can occur in two ways: either as an expression of an occasional mood, or as a habitual pattern of a character who preferably uses one or the other, downward curved or upward stretched.

Expression of same mood but different bodily carriages

In the art of dancing it is usual to develop the awareness and feeling for typical body carriage. As one discerns whether the pitch of a singer's voice is that of a bass, baritone or tenor, or an alto, mezzo-soprano or soprano, so one can detect in the dancer a natural tendency for rising or lowering movements. Such differences cannot be explained exclusively on psychological or physiological grounds. It is a mystery of individual constitution whether a person is, in voice or gesture, a "deep-mover", a "medium-mover" or a "high-mover". Classical ballet, the heir of the court ceremonies of the last three centuries, is an art form in which high-dancing prevails, yet it is astounding how many ballet dancers have no real high-dancing capacity and this disability makes their evolutions unnaturally cramped and sometimes ridiculous. If the deep-dance types would move in a deep-dance style, such as are seen in certain primitive or folk dances, they would be more at ease, since they would express what nature has predestined them to express. The criterion of a high-dancer, and indeed of any dancer showing a characteristic use of space, can be quite well recognised and described. A high-dancer has the natural tendency of uplifting and rising. He will prefer movements which stress the erectness of bodily carriage and, since acting against gravity, they will always show some tension. He will be able to convey an impression of utmost lightness not only in his leaps but also when his gestures are directed downwards leading to positions near the ground. The contrasting type is the deep-dancer who prefers to stress activity of the centre of gravity. Such a dancer might have a tendency to stamp and crouch, in which case the

In movement, as in voice, people show a constitutional tendency: to be high, medium or deep movers

Criterion of high-dancer

Deep-dancer

thrusts towards the floor are rhythmically pronounced, and the body carriage mostly curved. It is obvious that in drama the high-mover will best represent heroes, priests or beings of an elevated spirit, while the deep-mover will be more adapted to roles of earth-bound characters. The high-dancer will have the inner urge to assume shapes of great clarity and precision, while the deep-dancer will show a strong sympathy for rhythmic expression in which the formal element is secondary.

Mental and emotional qualities involved

It must be noticed that the dancer's typical trend towards these contrasting movement expressions does not depend only on the slenderness or height of the body. Although it is true that tall, slender people will generally be more inclined to use the high-dance, and short, stockily built people the deep-dance style, it cannot be said that this predilection will always correspond to their natural talent. Svelte people might be better in deep dances, and stout people in high. Mental and emotional qualities play as great a part in the determination of an upward or downward tendency as does bodily structure. It can be empirically established for which movement stress a dancer or actor is particularly gifted. Inherited or acquired effort habits too might have a great influence on individual dispositions of this kind.

Medium-dancer

The preference for height or depth is, however, not all that can be noticed in a dancer. Some technical capacities, such as that of agile turning or pirouetting, characterise a type which is seldom a confirmed deep- or high-dancing type. Again such agility in turning and pirouetting does not depend on bodily structure, but on inner constitutional factors difficult to summarise. Ease in turning demands a combination of quickness and balance, of formal clarity and abandon, and it can be empirically recognised though its sources cannot be fully explained. Turning always takes place at more or less horizontal medium levels of space, neither really high nor deep. Turning can be done either in an inward or outward direction, with gestures closing towards or opening away from the body. There are many people who do not show pronounced characteristics of a high- or of a deep-mover. Such people usually enjoy turning. They rarely have the crispness of a high-mover or the solid earthiness of the deep-mover. They excel in freely flowing, lilting movements which seem to surge like waves from the centre of their body into space and then recede again into stillness.

Significance of spatial directions in movement and bodily carriage

The movement directions, forwards and backwards, can be significant in many ways. The erect carriage of the high-dancer has frequently a dignified expression, stressed by a slight reclining of the head and by pointing the nose in the air; it can, however, assume an arrogant expression. The forward bent carriage of the deep-dancer could become aggressive or servile. The movement directions to one side of the body or the other stress its left and right symmetry. A sideways closing movement

can be used to express fear or shyness, whereas a sideways opening movement may denote pride or self-assurance. Mixtures of these extreme
spatial differences carried out by different parts of the body allow for a
rich scale of movement expression.

There is, moreover, the expressiveness of the path of movement. The *Symmetric and asymmetric*
path of a movement either on the ground, as a floor pattern of the steps, *positions and paths of*
or in the air, as the track made by an arm or leg gesture, can be straight *movement*
or curved. It can also be symmetrical or asymmetrical. Symmetrical positions and paths of movement are easier to understand and survey than the
almost incalculable number of asymmetric positions and paths. Symmetry
of movement is less passionate than asymmetry; symmetry hides while
asymmetry reveals personal excitement. The formal and bound character
of symmetric positions and paths of movement reminds one of the solemn
architectural beauty of a Greek temple. Movements expressive of religious
or ceremonial dignity will mostly be performed in symmetrical form. The
emotional discharge of unbalanced inner attitudes occasioned by passion
and exuberance will, however, be more appropriately expressed by asymmetric movements. Asymmetric movements are apt to degenerate into
unbalanced exaggeration, or a chaotic play of fanciful movement, and are
therefore less easily mastered and controlled; symmetric movements
offer a comparatively stable foundation for all expressive attitudes.
However, there are schemes of order within the asymmetric movements of
man. Several such devices have been developed by dancing masters in
different ages, and all try to master the asymmetry of movement expression. The subdivision of space into basic and derived directions is such
a device.* One could even say that the various methods of movement
training elaborated over the centuries have the fundamental aim of
mastering the disequilibrium of the body in asymmetric movements.

It is important to realise that the patterns traced by steps on the floor *Differences between*
and those traced by gestures in the air represent aspects of shape which *spatial patterns of two-*
are fundamentally different in spite of their similarity. While floor *and three-dimensional*
patterns are restricted to combinations of straight and curved lines *nature*
producing designs which consist of angular, round and figure 8 shapes,
these in the air are outlines of bodies in space. These are created by the
tracks which the different articulations of our body draw in the air
simultaneously. Their configurations may be coiled or twisted, they may
be intertwined in knot-like manner, or they may be straightened out and
move through space like a projectile. Whatever they are their shapes are
as pregnant with significance as are the patterns traced on the floor.

As an exercise it will be of value if the student tries out the various
shapes mentioned, and tries to find out how they fit into period styles. It

* See R. Laban, *Choreutics*, Macdonald & Evans, 1966.

is also valuable if the actor acquaints himself with period forms of working, travelling, eating, sleeping, greeting, and other social customs that have prevailed at various periods and in different countries. In performing the exercises the following points should be made clear.

(a) Meaning of movement going into the different directions of space.
(b) Expressions such as hilarity or anger when upward-stretched or downward-curved.
(c) Whether characteristics are those of a high-, medium- or deep-mover.
(d) Opening or closing turns.
(e) Symmetric or asymmetric movements.
(f) Nature of floor patterns.
(g) Shapes of arm and leg gestures.

An individual's development of movement awareness parallel to that of mankind: both progress spiral-like

The study of the history of human behaviour suggests a certain parallel between the development of the feel of movement during the life history of the individual, and the progressive refinement of movement knowledge during the history of mankind. The development from the first impulsive jerks of the body characteristic of infancy up to the stylised mastery of movement used by the adolescent can be compared with the development of primeval dances into those of later times. The line of progress in both the development of the child's movement, and the movement habits of communities in dance, is not unbroken. In both a spiral-like recurrence of more primitive styles is clearly evident. After each discovery of a new effort combination which for a period is cherished as the perfection of movement habit, a temporary return to more primitive forms sets in, because it is realised that specialisation in a restricted number of effort qualities has its dangers. If fine touch or lightness is exclusively cultivated, the strength of the body degenerates. If large and flexible movements are out of fashion, stiffness sets in. If extreme sustainment in behaviour and dancing is exclusively preferred, the capacity for quick decision and action is impaired. The same is the case when in such an epoch as ours speed becomes the ideal. The over-indulgence in the movement element of quickness generates nervous trouble. Too great flexibility shown in tortuous movements, as at some periods of history, creates a fussiness inimical to any straightforward approach to things and ideas. Snake-like mobility might induce avoidance of clear and direct issues. Social and individual brutality is often linked with the glorification of physical strength.

Changing fashions in social dancing show tendency towards establishing effort balance

Preference for a few effort combinations only results in a lack of effort balance. This lack is not always consciously noticed, but people, tiring of a movement fashion, try to introduce new forms which more often than not contain qualities of movements sharply contrasting with those previously used. New dances and new ideas of behaviour arise by a process of compensation in which a more or less conscious attempt is made to regain the use of lost or neglected effort patterns. This tendency can be observed in our own time. The European forms of present-day ballroom dancing have their source in the irruption of primitive dance-moods which counteracted our traditional stiffness. From the maxixe to the fox-trot, the Charleston, the rumba and the jitterbug, an increasing frenzy of movement seems to have invaded European ballrooms, yet, each time, such an excessively mobile effort display is tamed and reduced to quieter, even stiffened form.

Further back in history all kinds of exotic dances have been introduced when former movement fashions became too stale and lopsided. Very often such changes in effort moods have followed historical events, for example, wars of conquest, or have been introduced by explorers who have seen them performed by primitive peoples. Moorish dances followed hard upon the Crusades, and Red Indian dances the discovery of America. More recently we have seen introduced into traditional ballet folk-dance steps and movement forms of predominantly Slav origin. Few perhaps know that the Napoleonic wars led to the fashion of dancing on the tips of the toes, or rather the points of the shoes. This was a feature of the dances of Tcherkess warriors which was witnessed by French armies during the invasion of Russia. The introduction of this almost acrobatic feature rescued the classical ballet from the danger of sickly sentimentality, and from a meaningless display of softly interwoven linear embellishments. The stiff tension of the dancer induced by the acrobatic feature of the toe dancing was later corrected by the introduction of more impulsive movements as seen in so-called modern ballet.

Changes in effort moods often followed historical events

Origin of dancing on points in classical ballet

What frequently escapes attention is that these changes do not originate from the artistic or commercial whims of clever dance instructors, but are to be considered as real compensations, answering the deep-rooted need to keep alive the effort balance within the widespread movement habit of social dance. During both the life history of an individual and of mankind, continual progress is made towards the mastery of effort balance. This mastery is reached by degrees, yet, again and again, infatuation with newly-discovered movement possibilities leads to fashions and period styles in which the lopsided preference for a few effort qualities endangers their well-balanced use.

Mastery of effort balance endangered by movement infatuations

The student should try to invent scenes in which his knowledge of the different causes of movement stylisation finds practical application. These scenes can be serious or just caricatures, and should be performed, exercised, analysed, and attributed to specifically imagined characters. Such scenes can be selected at random. Imagination works freely and sometimes it would seem irrationally. The student should not hesitate to

Introduction to inventing and performing dance-mime scenes

follow even the most bizarre flashes of his imagination, only he should try to organise them so as to give a clear representation of character, values and situation, to the end that each sequence of mime scene becomes finally a well-shaped whole. Some of these flashes of imagination will have a dream-like character, others will be strongly realistic. The logical happenings of everyday life can occasionally be enhanced as in melodrama. Here, stress will be laid on the emotional content, and the happenings will have a more than ordinary character. In melodrama one exciting happening follows another, whereas in everyday life exciting happenings are sparingly dispensed. Happenings in dream states are often very vivid, even fantastic, but in everyday life, even with melodramatic enhancement, events and actions are rarely so intensely experienced. It is obvious that the movement style of a comparatively eventless everyday scene will differ from the dancelike one of melodrama. While in the first case shadow moves might prevail, in the second, larger and more active movements will be used. No rule can be established for behaviour in the dreamland of theatrical imagination.

In the following scenes some simple situations are indicated. It is for the reader to make them interesting through movement expression. Characteristic bodily actions (see Chapters 2 and 3) have to be invented and the value contents of the scenes should be analysed and repeatedly performed. Scenes in which several people are involved can be performed by the reader if he imagines the co-actors' movements; and each part can be mimed one after the other. Some examples of characters involved in certain events and situations might stimulate the imagination of the reader to invent his own mime scenes.

(1) Imagine a father sitting side by side with his daughter, both reading. There is a knock at the door. The daughter answers the knock and a man enters in an exhausted state. He is in fact an escaped convict, and collapses on the floor. The daughter brings water while the father goes to the phone to call the police. The daughter, feeling pity for the man, pleads with her father, and dissuades him from phoning. The intruder has just risen to his feet when there is a second knock at the door. The daughter turns to the prisoner with the intention of secreting him into an inner room, when to her horror he dies. The father answers the door. The police have arrived.

(2) Another theme: a tramp falls asleep by the roadside. He dreams that he inhabits a palace. He is in a room hung with rich tapestry and furnished with exquisite taste. Awestruck, he ignores the artistic treasures, and walks to a long table spread with food. Hunger gnaws him, and satisfied that he is alone, he helps himself wolfishly to the victuals. Suddenly the marble statues in the room spring to life and advance menacingly towards him. Then he awakens to find himself still by the

roadside, thankful that he is just a poor tramp.

(3) A theme of melodramatic character might show courtiers assembled in the throne room of a royal palace. The king enters with his entourage. He confers the accolade on two of his subjects. A slave enters and prostrates himself at the king's feet. He is closely followed by a huge brute of a man carrying a whip. The courtiers surround both intruders and proceed to drag them away. The king orders them to desist. Approaching the man with the whip, he takes it from him, breaks it, and orders his arrest. The king bids the kneeling slave to rise and depart, a free man. The court ceremony proceeds.

(4) A theme embodying everyday actions coupled with frightening day-dream depicts a servant dusting a room. While admiring a vase, she drops it and it breaks into pieces. Horrified, she looks about her to see if there was anyone who could have heard the crash and then collects and hides the fragments. In the suspense that follows she imagines herself in court, the judge pointing at her. Three men representing her master in three different forms are on one side, and three women representing her mistress on the other. A crowd of people surround and loudly accuse her. Suddenly the spell is broken. Her mistress enters the room and instantly perceives that something is wrong. The servant runs to her mistress, shows her the fragments and begs forgiveness, which is graciously given.

(5) Everyday commonplace action will be found in the following theme. A junior clerk arrives at the office. He is brought a pile of forms, upon which, one by one, he must write the same words. Wearily, he stops just at the moment when a senior clerk enters and leads him to a table piled high with books and papers to be sorted. While the junior clerk is engaged in this monotonous task, the employer enters the room. Greatly daring, the clerk summons up sufficient courage to complain about the senselessness of his work, only to be peremptorily ordered to do as he is told or take a week's notice.

(6) Of a similar kind is the next theme. A queue of people is lined up at a depot for the receipt of food parcels. The last parcel is given out, and a woman and her three children are left without a parcel. Another woman who has received her parcel sees their distress, and is torn between pity for the starving children and the thought of her own sick husband. The husband perceives her anxiety. He divides the parcel, returns one half to his wife and hands the other half to the woman with the children, himself going without. One of the children runs to him, and offers him a little doll.

(7) The following scene will be more melodramatic. Several men enter a room in ones and twos. Together they plan a felony and then depart. One conspirator is left behind in possession of important documents. He

is deep in thought, when two detectives enter. Treacherously, he informs against his fellow conpirators and is rewarded. The detectives leave and again he is alone. He locks himself in and paces up and down the room in nervous agitation. Presently a knocking is heard at the door. He erects a barricade of chairs against it. It is ineffective; the conspirators break in just as their former companion shoots himself. A hasty search makes it clear that the documents are elsewhere. The conspirators decamp.

(8) A dancelike theme could start with a princess writing a letter to one of her male favourites. Her tutor arrives and announces a court deputation. Reluctantly she rises. The courtiers enter and humbly bow. The princess advances reluctantly towards them. She turns to her tutor and informs him of her desire to be alone; all leave. She wants to be alone to indulge in a vision where she sees herself dancing with her own court favourites, all of whom she knows are secretly in love with her; and in imagination to revel for a brief while in the care-free, not too strait-laced life of an ordinary pleasure-loving woman of the world.

(9) A theme of dreamlike character might show a lazy boy sitting at table. Four servants in turn (each to be represented as a different character) bring him dainty dishes to tempt his jaded appetite. He refuses every offer. The maidservants proceed to clear the table. The boy's father enters, and tries without success to persuade him to rise. The father paces up and down in desperation. Then he beckons a servant and orders him to call in the musicians. They arrive and begin to play. Enlivened by the music, the boy gets up and starts to dance, first with one maid and then with another. The father orders all the servants to join in. They become exhausted, but he makes them carry on. The appetite to the boy returns after this violent exercise and the servants have difficulty in satisfying it.

(10) Another dreamlike melodramatic theme could depict an egotistical tyrant entering a large hall. He orders some workmen to build a platform; others to re-paint the walls; still others to decorate the hall with flowering shrubs and trees. The number of workers increases, and all the while he drives them on to work faster and faster, until tiring of his insane whim he suddenly calls a halt. The work displeases him, and he orders that the hall shall be restored to its former state. That done the workmen depart and leave the egotist to his own reflections.

(11) A day-dream or vision is described in the following theme. Two sisters enter and bring a basket of fruit to their old grandmother. Having caressed her with jocular affection they depart, and the old lady is left alone. In reverie she sees herself dancing a swinging waltz in the ballroom of a country mansion thronged with people. A handsome young man approaches her and she, now rejuvenated, gets up to partner him in the next dance. The crowd vanishes, and they dance alone. At the close, the

young man kneels and is kissing her hand when suddenly her father breaks in upon them. The youth retires and the young girl is escorted by her father to her own room. The vision fades, and the grandmother is left to her thoughts.

(12) Everyday action and dream are mixed in the following scene. A penniless musician ends his playing in the street and enters a shop to sell his violin. Pocketing the purchase price, he leaves the shop in dismay, for he has parted with his one, well-loved possession. He wanders as in a dream through the streets. It is bitterly cold, so he enters another shop to purchase a cloak. Once more in the street, desolate and lonely, without his violin, he hands the newly purchased cloak to a scantily clothed beggar.

(13) The following theme could be performed with variations. A young girl prepares the table for her brother's return. The door bell rings. She answers the bell and returns with a telegram informing her that her brother has been killed. Hearing her mother's footsteps, she hides the telegram in her pocket and leaves the room. The mother enters in great excitement. Putting down her shopping-basket, she removes her cloak and adds a few finishing touches to the table. The daughter re-enters the room, and mother and daughter embrace. The mother is surprised by the daughter's manner, but believes it to be due to the excitement of her brother's return. Regarding herself in the mirror, the mother decides that she must dress herself for the occasion, so she retires. The father comes home from work, and the daughter immediately shows him the telegram. Mother comes down and greets father, and he breaks the news to her. Mother and daughter sorrowfully clear the table. A variation of the above scene might be where mother and daughter were, at first, apprehensive and ill at ease and the telegram brought glad news. The scene might then end as before with the clearing of the table, but with the family donning outdoor attire to meet the long-absent relative.

(14) A dream with realistic movement content might show a ragged beggar-woman, once handsome, but now shrivelled and bent, limping along the dark deserted streets of a large town and begging for bread. The woman sinks down beside a stone wall lit by a street lamp. She falls asleep, and her former life of wealth and consequence returns to her in dream. She and the prince (for she really is a princess) are in the court room of the palace with other noble lords and ladies. The prince like the princess is handsome and popular, and all delight in his bright wit and gracious demeanour; but he bestows over-much attention on one of the court ladies. The princess grows bitter with jealousy and cultivates the friendship of a certain wise and goodly nobleman, one of the prince's high officers. She is not at all careful to hide her feelings towards him. The nobleman feels embarrassed by the behaviour of the princess, and by

strict court etiquette lets her know that he is a loyal subject of the prince and not one to be inveigled into intrigue. The nobleman's studied rebukes sting the princess to fury. In her outraged pride and loneliness, she paces up and down in her apartment planning revenge. A scheme slowly shapes in her distorted mind. She grows suddenly still. A malicious smile plays over her face. Opening a secret cupboard in the wall, she takes out a small phial of poison and hides it in her dress. She re-seats herself, all her natural composure and proud elegance restored. Ringing loudly for her maid she orders her to bring lighted candles and flagons of wine. Presently the prince enters with his court entourage and advances to salute the princess whose wicked purpose is masked beneath a gracious, winning smile. The servants pour the wine and hand it round, and the princess is caught in the very act of pouring the poison into the nobleman's glass. She is banished from the court.

(15) A realistic mime containing an extraordinary event is shown in the following theme. An old lady sits knitting by the fire. The fire is low so she goes out to fetch coal. During her absence two burglars enter through the window. They proceed to collect things. There is a clock which both would like to have. They argue, and eventually start fighting for its possession. The old lady comes in with the coal, puts down the bucket and stands between the burglars. Each pleads for possession of the clock. Somehow she pacifies them, and then they sit and consume the fruit she has offered them. Unobserved she mixes a sleeping draught and gives it to them to drink. The burglars fall asleep, and then the old woman telephones for the police. She mends the fire, and sits down beside it. Once more she starts knitting. The police arrive and march the two burglars away. The old lady is left knitting.

THE PRODUCTION OF GROUP SCENES

Art of movement in the theatre also requires corporate work of a team of artists

Beside the individual effort of the actor-dancer, there is another feature characteristic of the art of movement, and that is the team aspect, the corporate effort necessary to all theatrical work. Drama, Opera and Ballet are each the corporate work of actors, singers, dancers and musicians assembled together as a team to present a work of art in the form of a theatrical performance. Poets, composers and choreographers provide the idea around which these works of art crystallise. Directors and conductors supervise and assist in the creation of a common atmosphere which is enhanced by decor, costume and lights. The public, the counterpart to the stage, understands and enjoys the corporate effort of the performing company as a whole, and this whole has a definite form and character. This means that each one, the actor, dancer, singer and musician, must adapt his individual acts of movement, including the movements in voice-production, to the conditions of corporate creation. Through both his individual and corporate effort his aim is to touch the public, which is an essential partner in any performance. The gift, or the acquired skill, with which the performer produces his acts of movement must be augmented by the ability to adapt his skill to demands extraneous to himself. The extraneous conditions are the text of the play including the stage directions, the decor in which the performer has to play, the team of actors with whom he is associated, the guidance given by the director, and last but not least, the spirit of the audience to whom the performance is presented. All these factors, including the period style of a play, ballet, or opera, can be as alien to the performer as the character of the person he is to represent. The performer has to submit himself to the exigencies of the corporate effort and its unified style.

The director can arrange various groupings expressing the behaviour of people in definite situations. It is advantageous to have a few associates with whom to enact the following mime-scenes. It is also an excellent exercise to try to convey to the group what you want them to do by gestures only; gestures unaccompanied by words. Movement themes which can be expressed in this manner are as follows:

Exercises in group movement and corporate effort

(a) The gathering of a crowd on the stage in different but always corporate moods, everybody being gay, or triste, or suspicious, etc. Invent a common action presupposing the chosen corporate mood.

(b) Some simple common action is performed such as: working, playing a drawing-room game, sitting, standing, or just lolling about. Common action does not mean that everybody goes through the same movement. On the contrary, several different movements are harmonised into a common action.

(c) A group reaction to the following challenges: rushing, either amused or angry, to one corner, cautiously approaching a possible danger and shaping for a fight, expressing fear or indicating flight. The challenge can be given by one person or by a part of the crowd.

Compare the co-ordination of the different effort qualities of several persons moving together. The group movement will have a shape, a floor pattern, or a general direction or tendency in space.

(d) Dividing into two hostile groups, in growing disagreement or indignation one group exits and the other becomes calmer or more excited. The first group re-enters and reconciliation or actual conflict follows. One or the other group is overcome, and becomes resigned or capitulates.

A challenging soloist might replace one of the groups, and then the play of values might become more obvious than in the former group scenes. It might be shown, for instance, that some values are unattainable. The inveterate coward is ever a stranger to courage. Yet there is such a thing as the courage of despair, and then a miracle may happen. There are, however, many values open to the coward which might compensate for his lack of courage. Such values are: reflection, self-discipline, and the sense to hide his cowardice. Similarly, the dull and indolent person will rarely become energetic and fully active without some extraordinary stimulus. Patience and kindness could more easily change to anger, than dullness to energising intelligence.

Try to represent such developments of the struggle for values in a movement scene. Seek to establish the most appropriate effort sequences for every member of the group.

(e) Murder is rightly considered the cardinal crime; but its factual contrast, respect for the life of others, is by no means the highest virtue. The values of love, friendship and trustfulness are much more highly valued than simple repect for life. Similarly, respect for property is less esteemed than personal benevolence. Yet theft is considered as more abject than malevolence. Acts inimical to the material interests of another are regarded as ignominious, infamous and revolting, but to safeguard other people's material interest seems to us no more than decent. Offences against moral values, say love or friendship, are considered much more tolerantly. They are faults, but not abject faults. Acts expressive of real friendship and love, since they have a liberating and inspiring character, are exalted as the highest values.

Invent group scenes in which these general opinions about values are mirrored.

(f) A group can depict corporate expression of exaggerated gaiety or excitement with disappointment as its aftermath. Disappointment might be caused by the curious fact that objects of material value are divisible and distributable, and the more recipients there are, the less each gets. This contrasts strangely with the fact that moral values are indivisible, and yet their sum increases with the number of those who exhibit them. Consider the conception of love universal. The joy derived from the receipt of material goods and the envy aroused when they are withheld contrasts with the joy of sharing moral and spiritual values.

Invent group scenes in which material enjoyment or spiritual elation is depicted, and their consequences revealed.

(g) Show group reaction to surroundings, such as the solemnity one feels at a religious meeting; shivering in the cold; resistance to law and order; listening to a lecture, boring or interesting; relief from suppression; panic in danger; greed at a distribution of material goods.

Express in group scenes how social conflicts are based on material values and the desire to get as much as possible of them for ourselves;

whereas social unity has its roots in spiritual values which are indivisible and increase in sum with the number of those who share them.

Mime provides the occasion for expressing mental and spiritual concepts in movement scenes. The average actor may perhaps be reluctant to recognise this, and fatuous intellectuals of the now moribund Bloomsbury school would even have denied it. Knowledge of effort expression and the world of values enable us to prove to ourselves and others that the language of movement is not restricted to the representation of physical feats.

How can emotional and moral values as well as physical feats be manifested through language of movement?

It is recommended first to select elaborate forms of common physical action (without props) such as: building a dam or dyke; lumbering in a forest; aiding the wounded; a surgeon performing an operation; cooks preparing a meal; servants at their housework; groups performing military or athletic exercises.

When help is accorded to others, an emotional undertone will creep in. How does it manifest itself in terms of movement? The two satisfactions accompanying the attainment of material values and moral values are different. The first might be intensive, but somehow it remains shallow; the second runs far deeper. It is the innermost portion of the human being which responds to moral values. The profound equanimity and inner balance of a person habitually concentrated on moral values contrasts with the restless unsteadiness of those on the hunt for material goods. Individually, of course, everybody is free to prefer a full purse to the balanced depths of personality, but what of the choice? It is just this difference in tastes that makes the struggle for values, and therefore dramatic conflicts, possible. How are equanimity and balance mirrored in movement? It will be understood that mere verbal explanation fails to describe exactly the expression of these inner qualities. Although such inner experience is difficult to describe in words, it can be perfectly expressed in movement. The virtue of the theatre resides to a great extent in this fact.

The routine of theatrical performance lends itself to an astonishing number of movement observations. The stage manager and the performers, in various moods, reveal interesting forms of behaviour not only during the performance but also before the rising and after the falling of the curtain. The actor who is too early or too late in the wings; or who, half dressed, has forgotten some of his props; the actor who must be held back or urged on by the manager; the all-too-punctilious, or the apparently indifferent, or the nervous performer; and the actor so afflicted by stage fright that it induces nausea and fainting. All these are excellent models of movement expression. So too, the performer absorbed in his role who behaves unnaturally in the character of the role for which

Topics for study of movement expression

he is cast, the rivals in a play who start a quarrel in the wings, and the performer needing encouragement before he can proceed are all good objects of study for the mime. Observe the actor who enters the wrong way, or cannot mount the stairs, who falls, trips, stumbles in the wings or on the stage, and who is concerned whether he looks all right. These commonplace misfortunes, accidents and vanities result from, or are accompanied by, inner human attitudes. What is the difference between a rough, knockabout performance of such scenes, and the finer representation of suffering arising from them? Psychologically it might be said that the feeling of compassion lacking in the rough is clearly present in the finer performance. But do knockabout and slapstick scenes not frequently evoke in stronger fashion a compassionate feeling than the more serious melodrama? And which of these presentations is the more ridiculous?

Human actions and reactions observed provide material for movement scenes depicting tragi-comedy of existence

These are things of great interest in mime. Here are people who are comic in all seriousness: the performer absorbed in private business; chatting; flirting; swaying intoxicated in the wings; performers on strike; performers who criticise the lighting, decor, production and everything else except their own acting; performers who are over-eager to answer calls, who storm into dressing rooms, rearrange costumes, repair make-up, cause disturbance in group finales on the stage, try to force calls, run on the apron, and retire to backstage. What efforts and values are behind all these happenings? Observe the actors answering curtain calls, some keeping in character, others changing into personal moods of exaggerated friendliness or self-assertion, and others again showing too much mobility or stiffness. Observe the dejected performer discontented with his acting and his lack of success; the behaviour of the cast; of the public; the quarrelling between colleagues and with the manager, or the lighting personnel, or anyone else they can find to quarrel with; the vain performer too pleased with himself and his performance; the jealous actor, animated by professional or private feud; the more than jovial performer; embracing and osculating indiscriminately; the performer missing his calls, or inconsiderately pushing himself to the forefront; the all-too discreet performer; and the sentimentalist weeping copiously after the curtain is down. Many consequential scenes could be built up from such material. The reactions of the stage manager to such scenes and the controversies between him and the performers can have a serious, comic or burlesque character.

Human effort and movements are the sole means of action; they produce situations, and are the vehicle of change in life and drama. These happenings are as countless as the stars in the Milky Way. Similar motives may occur and recur again and again, but they all have their prototype in a kind of myth comprising the manifoldness of the basic tragi-comedy of existence.

SYMBOLIC SCENES

A grandiose plot could be understood as the conflict of two great mysteries: Life and Matter. Life can be shown as actively engaged with weight, space, time and flow and Matter as passively subject to them. Life lifts matter out of its heaviness. The first stir of matter, galvanised by the impact of life, is the source of all the unsuspected adventures and sufferings of mankind which history relates, and which dramatists use in their art. No living being can escape the mirroring of the fundamental motive in his individual life. To be conceived and born, to grow in body and mind, and to mature and to die is the great comprehensive movement of the passion-play which is everybody's fate. The essential means of human expression, bodily movement, follows therefore according to the fundamental scheme of life and existence. Each movement is conceived and born, grows and shrinks, and finally fades away into the past and nothingness.

The essential means of human expression, bodily movement, reflects the great rhythm of existence

It is a fundamental truth that the apparent cruelty of this primary motive of dramatic action is softened by a smile which imparts to life a fundamental kindness. We are moved to laughter, but it is compounded of tragedy and comedy because in laughing at our fate, we are impelled to the conclusion that man is by no means as important as he likes to fancy he is. We perceive with dismay that the same movements by which our highest aspirations are expressed also serve sometimes the most degrading ends. To shed tears in tragic moments of life seems to many of us ridiculous and our laughter in comic situations is frequently sinister, because it reminds us that we are but atoms in the vast universe. The materialist is inclined to consider his own existence as well as his efforts and movement as of little account in the whirling starry universe; that he is but a speck of matter set in motion by accident. The materialist is wrong. Life with its beneficent smile reminds us also that the apparent cruelty of existence makes us more careful and self-reliant in all kinds of situations, so that the apparent cruelty seems somehow to work for good, and give good hope to the idealists of our race. Man's material body is like an anvil on which the blows of life incessantly beat. An unknown blacksmith seems to be busy forging an as yet unrecognisable masterpiece. In this mysterious operation, the means of character formation, tears and laughter fly like sparks, their flashes illuminating the dim light of the smithy. Man sees then the vague outline of a being whom he can recognise as the Smith, and has a vague intimation of the masterpiece he is yet to be.

Such intimations sometimes come to us in the theatre, but we cannot fail to recognise that what we really see and hear there are the working movements of the actors, singers, dancers and musicians. The crude nakedness of the movements is transcended by an imaginative effort on the part of a corporate group of artists, and it is this effort, springing from the depth of their personalities, which conveys to us this intimation. The imaginative transfiguration of an effort from which movement streams spontaneously is indispenable to our liberation from everyday drudgery and to the recreation of spirit which we seek in theatrical performances. We want to see life from a new angle and thus more fully understand the hammering beat of our own hearts. Immersed as we are in the common outlook of the crowd, we want to forget our individual worries and interests, and long to laugh, or, maybe, to weep. Eager questions crop up in our hearts. Our desires can be fulfilled and our questions answered by a group of artists, individuals who have learnt to master their personal effort-chemistry and to co-ordinate this with that of the group. This cannot be achieved merely by mechanical discipline. Their bodies, of course, must be trained to the task so that they can create shapes and shades of imaginary action, but their creations must symbolise and be pregnant with a life beneath and beyond sense perception.

Penetration into the deeper layers of consciousness connected with the feel of movement

It seems to be a mystery how the senses, the eyes and ears of the spectators, can be made to feel what is beneath and above their sense impressions. A good company of actors, singers or dancers undoubtedly succeeds in making the spectator understand more than he actually sees or hears. Yet this becomes less surprising if we realise that penetration into the mysteries of life is intimately connected with the acquisition of well-defined qualities of the feel of movement. Curiously enough, there is a profound similarity between the process by which skilled movements are produced and the process of artistic creation. The common use of the word "art" for skilled action in any kind of human work reveals a secret desire of man. But skilled action is not always linked to the deeper levels of life. Sheer virtuosity, although comparatively speaking rare and therefore of a certain value, remains often an external excellence. Stage work requires more than purely mechanical skill. The harmonious flow of movement in gesture, voice and speech is governed by laws which become manifest in that strange world of hidden values revealed through tears and laughter in the theatre. An artist can display virtuosity for the satisfaction of his egotistic desires without bothering much about the degree of inner

Artist's display of virtuosity and his capacity of stirring up awe of life

truthfulness which the shapes of his movement may reveal. This diminishes his capacity to stir up the awe of life deeply hidden in the spectator. It is true that the virtuoso does not care much for the spectator's awe of life. He is mostly satisfied if he can arouse sufficent admiration to allow him to extract wealth and power from it. He has no heart-ache because, in blinding people with his sparkling gifts, he may hide the import of life rather than reveal it. He does not even notice that he withholds the spiritual food for which people crave. He is well content to satisfy their more superficial tastes, and is highly satisfied with external success. The assuagement of the more profound craving of the public demands an attitude and refinement of effort totally different from that partially offered by the virtuoso. Too many people today still think that all movement tuition and training is to exhibit the empty brilliance so highly praised in the virtuoso. This shows that most people have not the faintest conception of what the shape and rhythm of stage movement really is, and they are not aware of the deeper purpose it is meant to serve.

Movement is a process by which a living being is enabled to satisfy an immense range of external and inner needs. The actor-singer-dancer produces movements on the stage resulting in gestures, in sound and in speech. The repertory of movement embraces the whole range of actions by which everyday needs are satisfied; yet there is a marked difference between everyday behaviour and that exhibited on the stage. It is but a half-truth to say that life is reality and stage performances just make-believe. If make-believe means that the actor tries to create in the mind of the spectator some kind of belief in the deeper meaning of life hidden beneath external appearances, the statement is true. But one could not agree that it is an actor's business to induce the spectator to believe that his performance is just a faked copy of the events of everyday life. This, fortunately, is impossible because events, as well as movements, have to be carefully selected and composed into a whole, if an effective work of stage-craft is to be built up. Moreover, the selection and ordering must be devised in a special manner, so that a definite state of mind is evoked in the spectator and he is made aware of some particular aspect of the significance of life. Then and only then is it possible, with the help of the play and the movements of which its performance consists, to plant in the spectator's heart a seed from which a flower of the inner garden may grow. It may even happen that fruit will succeed the flower; fruit that takes the form of a new attitude towards life. The power to make people believe in such almost ineffable things resides entirely in the artist's well-cultivated movement capacity.

An artist requires a well-cultivated language of movement in order to touch his audience and give spiritual re-creation

CHAPTER 7

Three Mime Plays

The written text of a play offers the actor a solid foundation on which to work. It is true that the dialogue of a drama is not all that is needed. The actor has to translate the written word into audible and visible movement. The playwright gives him indications as exact as possible about the characters, the situations, and the values for which the persons of the drama are striving. Sometimes also the playwright gives indications of actions and movements. These indications, however, are bound to remain sketchy and perfunctory compared with the exact words to be spoken.

The dancer usually has music on which to base his performance. He finds there a rhythm and a mood, and in some ballets also a synopsis of the principal happenings. In the arrangement of his steps, the dancer is guided mostly by a choreographer. The steps and movements of tradition-al ballet have names, and the dancer can thus get a relatively clear image of the bodily evolutions which he has to perform. The most subtle mime content is partly indicated in the synopsis of ballet. Dancers who create their own dances have to find music suitable for the movement theme they intend to perform if they do not prefer to develop this in silence, that is without accompaniment. They have then to invent the steps and movements according to the music chosen, or else they have to have music or sound especially composed which will suitably support their choreo-graphy. This is often the case in modern ballet.

In modern dance, that is in the contemporary form of the art of dance, a composition is not built up from a fixed set of movements or steps, but the conceptions of space elements and effort elements are a guide to formulating the actions of the body. For many modern dancers the "literary dance", a dance whose contents can be described in a synopsis, is not dance at all, but dance-mime. Movements familiar to the spectator through their similarity to those used in everyday actions and behaviour are suspect to these modern dancers. They believe that the real content of dances can never be described in words, and therefore they abhor dance movements which can be understood intellectually. In modern dance, stress is laid on those aspects of man's nature which are antecedent to speech and words, and which are outside the sphere of any personal recollection of real happenings or events in the normal struggle of life with its logically inherent emotional reactions. In modern painting the subject-matter is abstracted from ordinary sense impressions received from the surrounding world. In like manner the modern or abstract dancer attempts to convey an insight into a strange spiritual world of his own. It is assumed that this world constitutes "the primary heritage of embodied human spirit" as one of the modern dancers expresses it. Dance is for these artists a manifestation of those inner forces out of which the complications of human happenings grow. They are not interested in the situations, happenings and conflicts of practical life, not even in the enhanced form in which drama portrays such conflicts. They think that they can scoop deeper from the reality underlying the ordinary experience of life, and that dance is the language in which this deeper awareness can be conveyed to the spectator. The plot of all abstract dances consists solely of the development of a characteristic movement rhythm as it proceeds along certain paths in space. The impression gained by the spectator from such rhythmically progressing space forms or patterns drawn in the air by the human body is by no means merely an ornamental arabesque. As such it would have perhaps a decorative value of external beauty but no deeper significance.

Modern dancers, and the enthusiastic admirers of modern dance, see in these evolutions in space the messengers of ideas and emotions which surpass thoughts expressible in words. It is obvious that movements conveying such inner experiences must be carefully selected and assembled into rhythms adequate to the inner sensation. This selection is the result of deep concentration, and it might involve a strenuous effort taking time to make. The mental or intellectual factor is restricted to the memorising of the right series of movements which are the result of the selection. Sometimes a title is given to such an abstract dance; it does not, however, indicate any content, but solely a mood. An abstract dance can never have a synopsis because its content cannot be described in words. The whole idea is by no means so preposterous as it might seem to people who have never before paid much attention to the problem. In fact almost all traditional folk dances and national dances are abstract. Nobody can describe the content of say a "chaconne" or "gavotte". All that can be described is the sequence and the rhythm of the steps, and the gestures which the dancer has to perform. The modern dancer vindicates the right to create personal dance inventions built up from sequences of non-conventional evolutions of the body. He hopes to perform thereby a kind of magic invocation or incantation of life powers. These powers, or forces, have no names in our modern vocabulary, but we all know them when we meet them in the form of unaccountable inner drives, attitudes, emotions and beliefs. Modern man hides away the experience of such powers. He is probably ashamed or even afraid of them because they cannot be intellectually explained.

Some modern dancers try to adopt the terminology and the thought of contemporary psychological analysis in order to describe the contents of their creations. They meet the dilemma of using these terms either by employing them in their accepted scientific meaning, or by using them freely in an almost poetic or symbolic sense. In the first case, the content of dance is intellectualised in the same way as when using other words with conventional meaning. In the second case, it would be necessary for the artist to explain in detail the meaning of the terms used, which again is impossible. It seems that all attempts to speak about the deeply felt struggle of natural forces would only be possible in a community with a generally accepted religious or philosophical terminology. All such considerations have sometimes stimulated modern dancers to perform in silence, without musical accompaniment, and to let the language of movement speak for itself.

The mime-actor will find almost no written text on which he can base his work. Compared with the great number of plays accumulated for centuries in dramatic literature, and the equally large number of scores of dance music, one can say that mime literature is virtually non-existent. It is most difficult to unearth the few synopses of old mime plays scattered here and there in historical treatises on the theatre or in other books. No tradition exists about the composition of mime plays. An artist who wants to create such a play is left entirely to his own intuition and invention. The invention of short scenes of mime is not too difficult, and it is hoped that the examples sketched in previous chapters have given the reader the incentive to create such scenes. A longer mime play, consisting of several scenes or acts, may, however, contribute new viewpoints on the study of the art of movement on the stage. Therefore a few sketches of mime play are added here. They may offer the reader the opportunity of following up coherent effort developments and struggles after values over a wider field. Such sketches of mime plays are not works of art. They belong neither to poetry as dramatic literature, nor does their inherent rhythmicality entitle them to claim that kind of artistic value which a piece of music has. A good sketch of mime has the rather negative task of omitting all situations and ideas that cannot be represented by movement. The art form of written mime plays may arise if people learn to take the trouble to write and read mime scores noted in exact terms of movement.

Readers who have studied and assimilated the contents of the earlier chapters should be able to give an exact account of the movements which they use in interpreting a mime plot. The three verbal descriptions of mime plays given here are only in skeleton form; and it is the task of the mime-actor to fill them with a living contour of effort rhythms expressed in definite movements. The interpretation, no matter whether it is rich and imaginative or poor and dull, is the real act of creation. It goes far beyond the scope of the present publication to give a fully detailed description of possible interpretations of the synopses together with appropriate movement sequences. It is desirable for the reader to remain free to create his own version by finding the effort rhythms and bodily actions which would characterise the acting personality and his behaviour in the given situations.

The first mime sketch deals with the age-old theme of the "Magic Flute", best known in the form of Mozart's opera. Much has been written about the symbolic meaning of the story. It is considered to be an artistic transcription of an ancient initiation ritual. Such rituals have a common basic idea, although their outer forms vary with different races and in different epochs of history. The basic idea is the introduction of adolescents to the awareness of the overwhelming power of the life forces driving them towards love, ambition and the satisfaction of curiosity. The mime sketch presented here tells only the beginning of the story: how the youth Tamino gets his flute, and the first use he makes of it. Tamino's further struggle with all the elements of nature, and his final acquisition of a mature mentality, would be the content of later acts of the play which are not given here.

TAMINO'S QUEST

Tamino and his companions who acknowledge him as their leader in spite of his youth, come into a barren, rocky desolate land. They enter the unaccustomed surroundings in fearful and apprehensive mood.

In the dusk of a misty evening, a solitary bush on the edge of a rocky crag assumes the shape of a crouching giant. A storm is raging; its noise and that of the nearby sea seem to merge into a superhuman voice, chanting a mysterious spell. It is the voice of the giant, Monostatus, the servant of the Queen of the Night.

Monostatus has been sent out to waylay the most courageous of the young men, whom the Queen of the Night wants to use for her own secret purposes.

Tamino is unaware of this, but as he is the only one who stands steadfast in the face of the unknown power, he is chosen by Monostatus to meet the Queen of the Night.

After his companions have been hurled away out of sight by the storm, the incantation ceases and the deep blue firmament studded with stars appears high above him. It is the mantle of the Queen of the Night who beckons to the giant Monostatus and orders him to endow Tamino with the magic flute. The Queen remains unseen by Tamino; she has her own plan how to get in touch with him.

Tamino is exhausted and sinks on to the gnarled roots of the solitary bush; deep darkness envelops him. As he awakens in the early dawn, fog hides the bush; the giant Monostatus has disappeared. Tamino finds the magic flute lying at his side. He is confused and bewildered but tries to play it; the enchanting notes of this new treasure revive him. He is carried away with the beauty of its melodies, and he dances out of sight with gladness and joy.

The Queen of the Night appears again, but her majestic cloak dissolves in the morning mist. With artful cunning she assumes the appearance of an old woman clad in a grey tissue which her female attendants weave out of the fog. The Queen of the Night in the disguise of the old woman huddles down and waits for Tamino.

He reappears gaily playing on his flute. He sees the old woman and advances slowly, pitying her plight.

The Queen pretends to be startled by his sudden appearance, and feigns surprise at seeing such a fair youth wandering alone across the deserted highland. She seems to wonder what or whom he is seeking so far from the haunts of men.

Tamino knows only that he is seeking something, but what it is he does not know. The old woman is bitter. She stares miserably before her, knowing that what he seeks is something that she has lost. Tamino offers his help to find her lost treasure, whereupon the old woman draws out from the folds of her cloth a shining crystal. Tamino lifts the crystal to his eyes, admiring its rainbow-coloured glitter. Turning it towards the horizon aglow with the rosy light of the rising sun, Tamino sees in it the most beautiful vision he has ever met, the vision of a young girl. The tender clouds over the golden streaks between earth and heaven have assumed the shape of Pamina, the daughter of the Queen of the Night.

He is immediately filled with fiery passion and overcome with the longing to reach her. Forgetting the crystal, he climbs over the rocky land in an endeavour to touch the enchanting phantom. The crystal drops to the ground and breaks into a thousand pieces. The vision is gone.

In desperation Tamino implores the old woman to explain to him the nature of the vision, and to tell him where he can find the beautiful creature who has so touched his heart.

The old woman then rises to her full height and throws off her grey cloak of disguise. Spellbound, Tamino faces the majesty of the Queen of the Night. He understands that the maiden whom he saw is her daughter, Pamina. He implores her to lead him to her daughter, but she orders him to compose himself, to sit down and wait.

She withdraws to one corner of the rocky plain, while in the other corner there appears in the background the kingly figure of Sarastro, her husband, the Master of the Light. Pamina kneels at her mother's feet, whilst the Queen of the Night tries to hide her daughter's eyes so that she will not see her father, Sarastro. There is a feud between the Queen and Sarastro because each wishes to guide Pamina according to their own principles. The Queen of the Night, confident that her daughter favours her, calls upon Pamina to make her own decision as to whom she wishes to follow — her father, Sarastro, or her mother, the Queen.

Pamina, detaching herself from her mother's side, wavers uncertainly in doubt and despair, but the will of her father is stronger, and she seeks protection under his arm. Sarastro, guiding his daughter away, disappears.

The Queen shows Tamino how bitter her disappointment is and Tamino realises that the treasure which the old woman has lost is her daughter. He also knows that Pamina is the goal of his quest.

In sight of the frustrated and revengeful Queen, he wants to rush towards the spot where Sarastro and Pamina have disappeared. But the Queen holds him back, and hypnotised by her starry eyes he sinks to her feet as if lifeless.

The Queen beckons to Papageno, half bird, half man, who comes hopping, skipping, dancing round the corner of a rock whilst merrily playing on his pipe. The Queen leaves Tamino to the care of Papageno, who at once proceeds to awaken the sleeping youth. Tamino rubs his eyes as he sees the amazing birdman, and he thinks that he must be dreaming. Papageno bows gaily to him and Tamino courteously replies. Tamino wants to ascertain whether Papageno is really a living being, but the bird-man is too quick for Tamino. He darts away. Every time Tamino tries to touch him, he darts off in another direction. Eventually they dance together, their steps becoming quicker and quicker. Finally Papageno points towards the glowing streak of light on the horizon, and Tamino realises that the birdman is able to lead him towards Pamina. He is however unable to keep pace with him, and Papageno disappears. Tamino is left incapable of moving, as if some invisible barrier were preventing his advance.

Solemnly a bearded stranger, who looks like a high priest of an unknown cult, approaches. Tamino is doubtful of the man's intention, but soon he feels that the imposing apparition is trying to help him. The high priest points to a heavy door. It is the entrance to mysterious grounds in which Tamino is sure Pamina is hidden. In a solemn and powerful dance the high priest shows the difficulties and dangers awaiting any youth who sets out on the quest. All the elements are against him who attempts to enter the enchanted place. There is a girdle of fire, and a girdle of water to be traversed, and the ground is as shifting and treacherous as the clouds in the sky.

But Tamino stands resolutely facing the stranger. He is eager to continue his journey. The bearded man, seeing his strength of purpose, turns slowly and indicates with an ominous sweeping gesture the road to be taken through the now opened door.

Fired with enthusiasm, Tamino advances, but finds that the surroundings change continuously, as if in a whirling mist. Wherever he looks there is no set path, and he is unable to discern which way to go. He moves frantically to the right, to the left and round about within the shifting landscape. Just as he believes that all is lost, Papageno jumps up in front of him as if out of the turbulent air. The birdman leads him along a hardly distinguishable, winding path, first this way, then that, faster and faster and faster, round and round until he stops abruptly — and there sits Pamina, lost in the world of her dreams.

Tamino stands and stares at the vision which he recognises as the same that appeared to him in the crystal. He is enraptured by her beauty. She is far lovelier than he imagined possible. Hardly daring to breathe, he approaches her softly, but she gives no response. He approaches her again, but with her head bent low she makes only a simple slow gesture of refusal.

Tamino is bitterly disappointed that the moment for which he has fought and waited so long should prove to be so different from his eager expectation. He kneels to Pamina, he begs her, entreats her, implores her to look at him, to follow him, to accompany him back into his world. He has forgotten his purpose to bring the daughter back to her mother; all that he wants is to keep her for himself for ever.

Pamina in her deep distress caused by the feud of her parents looks away from the impetuous youth and shakes her head sadly.

Tamino loses his patience and becomes angry because Pamina does not return his feelings. Hardly knowing what he is doing, he invokes his first superhuman helper, the giant Monostatus, who appears at the head of a band of grim warriors. They march in strict formation, strong and powerful, bearing down upon Pamina to capture her and carry her away. Just as they are about to seize her, Pamina lifts her head and looks away with an enigmatic smile. Immediately from the direction of her gaze, Papageno the birdman hops in. Merrily playing his pipe, he dances round the awestruck soldiers. Anger and bitterness melt away in Tamino and the band.

The whole company is infected by the sparkling gaiety of the birdman's tune and dance, except the giant, Monostatus, who remains petrified in his sombre crampedness. Around this rock of desolation, everybody — Papageno, the warrior and even Pamina — dances to the vibrating rhythm.

Tamino, though amused, feels his attempt to abduct Pamina thwarted, and he resists the urge to join in the merry round led by Papageno. He puts his magic flute to his lips as if questioning the flute what he should do now. The tune of the flute melts with that of Papageno's pipe in a melody of gay exuberance, and Tamino, almost against his will, starts to approach the dancing Pamina with eager steps.

He does not succeed, however, in reaching the maiden, because the warriors and Papageno always get in his way. Monostatus has disappeared. Pamina, Papageno and the soldiers also vanish.

Tamino is once more left alone. He goes on dancing without being able to stop until he falls exhausted to the ground and loses his senses in a heavy sleep.

Such description should not be used as a synopsis destined for the spectator, but as a stimulus for the dancer or producer who has to find the appropriate expression of the story in movement.

For this purpose one could add a precise plan of floor patterns and effort sequences, or even an exact notation in kinetography of all bodily actions. The music indicating rhythm and mood of the various sections of the plot would add further stimulus to the production. It would be possible also to discuss the symbolic meaning of this "myth" or "legend" or "ritual" or whatever one might choose to call it. The reader is left to ponder over the symbolic meaning and its representation in movement.

THE DANCE OF THE SEVEN VEILS

The verbal description of the content of a mime or a mime-dance can take many different forms. Here are the notes of a young dancer concerning her dance of the seven veils of Salome. The story is known to everybody. Herod, the King of Judea, asks Salome to dance for him. She demands the head of the prophet, Jochanaan, as the price for her dancing. The prophet is beheaded, and the head is brought to her on a dish. The notes of the dancer describe the feelings from which her movements originate as a response to the whisper which seems to her to emanate from the lips of the dead. She heard the head of Jochanaan say:

"Salome! When I first saw your eyes, as they looked at me and rested on my body, I blushed. Not for shyness or shame, but because I was reminded of the animal I am. My self-attention long since lost was awakened.

"Salome, I see you become conscious of your body, your legs, your breasts, your lips — and you turn in vain. I see you all round, from all sides and from within your blood.

"Shame is the shadow of your self-attention. You would like to conceal your body, but where will you hide? You cannot endure to meet my gaze — and neither could I when I first saw you.

"Cast down your eyes and look askant as much as you like, or look at me with all the effrontery of a queen. Our eyes soon become restless, as did mine when I first saw you.

"Bury your face in your gown. Throw your whole body on the earth and hide your head.

I did not. I stood erect, but my heart beat eagerly. "I was covered with confusion, I stammered and made awkward movements and strange grimaces, I know. It is your turn now! Your bright eyes and your excited behaviour show me that the quietude of your mind is gone.

"The deprecatory remarks and the ridicule you addressed to me as I first saw you, and also the praise and the admiration to which you changed flatteringly, were not so touching as your intent gaze.

"Do you feel the gaze of my broken eyes from within your blood? You shudder, you would like to shake it off, to shake off my gaze — in vain.

"As your gaze was in my blood as I first saw you, so is mine now in your entrails.

"Take off the first veil of my head. It was my blushing.

"In hearing your flattering words and feeling your approach, your touch on my arm, my shyness changed into surprise, this into astonishment and this into amazement. My mind was not far from terror.

"But it was not your fingers which terrified me, but my own skin.

"Raise your arms further and further as I raised my eyebrows. Open the caressing curve of your body, as I opened my eyes and my mouth in amazement and fear. Swallow the news of my revelations into your lips, as I swallowed the dreadful perfume of your breast. Dumb and terrorised I was by myself. Dumb and terrorised you shall be by your fear.

"Swing your head to and fro and beat your breast as I did when I first saw you. Be immobile, incapable of moving a limb, as I became with my growing terror before life. Throw away the veil of my blushing and jump; jump high — but you cannot escape.

"I did not jump. I was fettered in iron chains. Only my heart jumped higher and higher. And I opened my eyes in increasing terror. Then I could see, see what it was that frightened me.

"My forehead was wrinkled. You asked me why. I could not answer. My eyes would fly away. My nostrils opened with my mouth. I needed more breath to escape the terror which came to me from the warmth of your body. The coldness of my limbs is creeping now over your flesh — you feel it?

"My muscles relaxed more and more. I feared to faint, to fall at your feet. My jaw dropped and my unheard cry of alarm suffocated in my throat.

"Do you feel my cold fingers around your neck? Cry, shout, howl — I could not do it — as I first saw you.

"And I raised my opened hands above my head and touched my own face to cover it in shame before myself. Trembling I fell on my knees — before you — as I never did before save when I was praying. And I prayed to myself to cease to feel, to hear and to see.

"Fly, fly away — if you can — I am everywhere. You are tortured and twisted through and through — creep nearer, nearer to my head and tear it away — the second veil, the veil of my horror.

"Contempt grew in myself against myself. A disgust as if I had swallowed tainted meat, smelled the dirtiest odour, touched the nauseous mass of putrified snakes, seen the horrible face of the nothing. A retching and vomiting shook my entrails, and I lifted my head, bending it backwards and closing my eyes.

"Show your strongest disdain, your queenly haughtiness, little Salome, and look down, down on my lonely head, here on the floor at your feet.

"But you cannot be so contemptuous of my skull as I was of myself and my intercourse with your warmth — as I first saw you. Snatch it away, push it away, press, press it away, the entwining knot which my invisible arms are tying around you. So felt I as — you remember, Salome? — you tried to break my haughtiness by hanging your soft arms on my neck. But it was not you, Salome, of whom I was contemptuous, but myself.

"My scorn and my contempt with which I pushed you away was the disgust of my own flesh. Do you feel, can you feel, the nausea before your own glittering body?

"You looked guilty with your look of stealth; but it was I who was the guilty one. My vanity and pride were monstrous. And this vanity and pride descended on you. Hanging my head over my secret I shrugged my shoulders. My restlessly moving eyes could not support your offended gaze.

"And you were sheltering in slyness; turning your head to the right and your eyes towards me to the left, you reflected upon means to break my pride. You were suspicious of my firmness. You took my shrugging as an acquiescence and — Salome — it was — submission.

"But raised and puffed up as a peacock or a turkey you tried to look down on me, but I was higher than you, if not haughtier.

"My neck contracted and I became helpless. I bent my back like a dog, awaiting patiently the blows of its master.

"You turned away and left me to my contempt and disgust for myself.

"Salome, raise the third veil, the veil of my broken pride.

"My hatred and anger turned against you. The injury my pride had suffered made my pulse throb. I blushed again. My chest heaved and my nostrils quivered.

"Are you raging against yourself that you have not understood my submission? Do you hate your own stupid haughtiness and lack of patience?

"I see your teeth clenched. I feel the vengeance of your insult to me in the tension of your muscles. Rigidity before the attack. Assault yourself, raise your arms, clench your fists, beat yourself, slay yourself. Dash yourself to the ground and beat the earth. Threaten the heavens and tear into pieces with your purposeless and frantic gestures everything within your reach.

"So did I in my violent rage when you left me, Salome. I was rolling on the ground, screaming, kicking, scratching and biting my own arms — instead of yours which are gone.

"But it was not passion for you, Salome. It was the passion of my offended pride.

"I trembled then, and my paralysed lips refused to utter my curses which I wished to send to you.

"My grinning became a violent laughter of hatred. My mind was darkened by the wish to kill you and myself. The brute nature which was awakened by your touch shook me convulsively. It is not you who is swaying yourself backward and forward, Salome. It is my will working in your fragile body, my wild hatred of you and myself.

"My protruding eyes stared in the remoteness hoping to reach you, to drink you in, to kill you with the glance of my dilated pupils.

"But nothing, nothing, only voidness and the rising twilight of the evening were visible in the loneliness of my den.

"Pull it away, the fourth veil, the veil of my rage and hatred, Salome.

"I regained my mind to find myself sitting with lowered eyebrows. Sit down, Salome, and think, think, think of our lost absorption in each other. Think of the eternal obstacle between us, think of our love-hatred.

"You are perplexed, afraid, haunted by the thoughts which dance in your head. They are the same thoughts which danced in mine as I recovered reason and began to meditate about what had happened.

"What had happened exactly? Nothing. A playing child has lighted a fire — the damage can be repaired. The fire can be extinguished.

"You are frowning, Salome. You don't believe that we can put out the fire? You cower, the frowning involves your whole body. You shrink, you disappear — Salome, Salome, stay, it is not yet ended — and will it ever end? — Think, Salome, think.

"You look as if you would strive to the utmost to distinguish some far distant things. What are they? Are they my thoughts which will meet with your thoughts?

"Search, Salome, search.

"You close your eyes. You do not wish to see the light. But inside, inside, Salome, where I am dwelling, I, Jochanaan, like a dragon in a cave — in your head, Salome, take it, hold it with your hand, press it, it will burst. Jochanaan, the dragon, will be set free, will rush to the heavens and leave you alone — alone on earth — alone with your sulkiness, your despair, your — murder.

"And I in my hatred wished to murder you, Salome. My eyes become vacant, empty. The pain within you raises your lower eyelids. You raise your hand to your forehead, mouth, throat. You fold your arms across your breasts. You are cold, Salome? It is the icy breeze of my whispering which makes you shiver. Ugly, ugly is life, is death. Ugly is birth and love. Showing a cold shoulder to it, turning away from it. Away with thoughts bitter and thoughts torturing, away with this swarm of flies.

"The fifth veil is the veil of my brooding soul.

"When I awakened this morning, Salome, I knew I loved you. The small strip of sunlight furtively touching the border of my den made me raise and stretch my arms towards the light. Glory filled my heart.

"Raise your arms, Salome, and dance! Dance around without purpose and without thinking why! This is joy, joy of life, joy of tenderness.

"Clap hands, stamp, and laugh. Children are always laughing when they play! Meaningless laughter is the best laughter — and so laughed I when I knew that I was loving you!

"Chuckling and giggling in mere happiness when feeling the supreme caress of life, this is the real sense of existence.

"I never knew it before, since my childhood I never laughed, what fool I was! And as a child I stood aside at the play of others I could not see. I could not understand why they laughed. But now I know it. You taught me to laugh. You liberated my rigid joints. I wished my fetters would fall from my wrists and ankles. Not because I wanted to be free, but because I wanted to dance.

"But is the shaking of your body not painful? Painful like the convulsions of dying? Life and death stand always near together, Salome, two inseparable twins — and you sent me the other.

"All this is of no avail, Salome, since the soles of your feet are tickled by the ground on which my head lies immobile, unshaken. Don't touch this terrifying soil which drinks my blood! Don't touch it! Your tender feet may become bloodstained. Jerk and curl your toes and the spasmodic heaving of your chest will cease. It is not fear, Salome. It is not the terror of my dead head which cramps your fingers. It is laughter! It is love!

"The delusions of your wealth, your rank, your grandeur, fall away from you as I wished that my fetters would have fallen. Your hips tremble in foreshadowing the leaps and flights to which they will give birth.

"Your bright and sparkling eyes shine tensely in your upright face. Don't bother about my closed eyes! But close yours also, Salome, in the last ecstasy.

"You have the desire to touch my hideous head. Approach, Salome. Approach, tenderly, softly. Your humble kneeling, your hands turned upward, and the joining of your palms show me your devotion. Come nearer, Salome, closer to me, much nearer, and take it away, the sixth veil, the veil of my love.

"After my paroxysm of joy I saw the two soldiers descending to my den. What may they bring? The freedom you promised me if I would kiss you — I have not done it — but I am eager to do it — have you been reading my heart?

"I never defended myself. I was too proud and life seemed to me worthless. But now as one of the soldiers drew his sword — I understood — and violently, frantically I fought for my life.

"But my eyes became dull in thinking of you, of your gruesome, insensitive youth, and my body lengthened for the last time as I wished to stretch myself into the light; out of my den, to see you, Salome, in one last glance.

"When you come to lift the last veil, you will see the wondering arc of my eyebrows. They ask you 'why?' They froze in this position encompassing the last vision of the tender arc of your breast. The waves of the wrinkles extending across the whole breadth of my forehead are the frozen waves of an ocean of love.

"Beat your half-closed hands together, rhythmically, without interruption, endlessly, as for ever. It is the beat of time which has ceased to travel through my heart.

"Work, work with your tightened muscles as long as you live. My muscles are tightened also, but I cannot work. I move no more.

"The corners of your mouth are depressed as in disgust, anxiety and dejection. Bend your neck. My proud princess, why are these wild jolts so different from your dance of joy, fluttering over the earth as leaves in the autumnal storms?

"Do you feel my despair? Not the despair to die, but the despair to see you — no more.

"Cry, Salome, and mix your tears with my blood. And the rarest flower on earth, real compassion, may grow in your heart. Bespatter the germ of the flower with our mixed stream of life, with your tears and my blood.

"And now, Salome, raise the last veil, the seventh veil, cautiously, slowly, the veil of my despair, and bury my dead lips between your bud-like breasts."

The dancer who desires to translate this poetic description of the emotional content into bodily actions has to consider the rhythm which may or may not correspond to the verbal rhythm. He has furthermore to consider the hints given concerning certain effort configurations. Indications of shapes in space are also contained in the text. However, the main thing will be to endeavour to catch the mood of each passage and to find the appropriate movement sequences according to personal taste.

The next mime is a fairly realistic happening.

In the director's logbook the main groupings and movements on the stage are described beside the action or mime content of each scene. The reader will notice that the description of the groupings and movements leaves freedom to the individual imagination of the director or actor who attempts to perform the mime.

It is, of course, possible to fix and to describe every single movement exactly.

THE GOLDEN SHAWL

Scene I

Groupings and Movements

1. The play opens with an immobile group of three persons. They turn their backs to the audience. The formation of the group, and the bodily tension of the individuals, suggest attention and expectation.

2. They remain immobile as the men enter, and perform working actions. The relatively quiet steps and actions of the workers build up a crescendo of movement, leading the eyes of the spectators from the immobility of the central group,

Action

1. When the curtain rises, three girls look out, immobile, towards the sea. One is a young, very shy girl.

2. Two fishermen enter, and one of them beckons to a third who drags a fishing net over his shoulder. He looks at the three girls, but, deciding not to approach them, he advances towards the other two men. They greet each other jovially,

over the sustainment of walking steps and working actions, to the alert quickness of some curious women who enter the stage a moment later.

The direct tension of the immobile group of girls is followed by the flexible but heavy gestures of the walkers.

and go out together with their fishing tackle.

3. This flexibility is enhanced by the alert lightness of searching curiosity shown by the entering women.

The floor pattern of the scene develops from the fixed spot where the immobile girls stand to the relatively simple and direct paths of the working persons, and later to the scattered criss-cross running steps of the curious women.

3. One of the waiting girls has seen something out on the sea. The other girls also peer forward to descry what she has seen. They are joined by a number of other women, the daughters or wives of the fishermen. All are excited because a strange boat is approaching the harbour. Several girls run out to discover who is in the boat. The others wait expectantly.

Two girls dash in excitedly. They have met a pedlar, obviously from foreign parts.

4. With the entrance of a strange figure the pattern concentrates into a half-circular group forming around him.

The immobile group of three girls has dissolved quietly. The shy girl removes herself to the background.

The new effort of expectation of the large group around the pedlar is lively, contrasting strongly with the stillness of expectation of the three girls.

4. The strange pedlar enters accompanied by girls. He is a high-spirited friendly sort of fellow, and he arouses great curiosity.

He shows some of his wares, gesticulating vivaciously as he does so, and the girls are full of admiration for what he shows them.

Notes to the First Scene. The features mentioned above are all movement happenings. Effort patterns change and follow one another, creating various shapes in individual gestures as well as in group formations. A few recognisable working movements are done, but no other conventional mime movements have yet been used.

Scene II

Groupings and Movements

1. The first movement of this scene consists of several short dances by individual girls, who move away from the group standing round the pedlar and finally return to their places. Each of these dances shows a special movement character expressed in corresponding bodily actions, shapes and rhythms. A general dance of the girls follows, in three groups, each around one of the three soloists of the first movement.

2. In the next movement the pedlar approaches the shy girl in the background, and engages in a long duet with her. The action content of these scenes is important for the composition and changing expression of the duet.

The character is indicated by the reluctance of the shy girl to move into the foreground and the half-contemptuous, half-encouraging support which the group of girls offer in accompanying evolutions.

3. The next movement is a solo action of the stranger which calls the attention of the whole girl-group (except the shy girl) to him. The girls gather closely around him.

Action

1. First one girl advances to purchase a belt from the pedlar, then another, this time to buy an ornament. A third approaches pertly to select a comb, then several girls flock round the pedlar to explore his wares. All dance, holding their purchases.

2. Suddenly the pedlar sees the shy little girl in the background. "Who is she?" he asks, but the girls make no answer. She is too insignificant to be noticed.

The pedlar advances, for his interest in her is aroused. The timid girl makes a few hesitating steps towards him, but falls back amongst the other girls who now try to encourage her to approach the pedlar.

3. The pedlar causes a general stir, searching mysteriously for something among his wares, and finally producing a wonderful golden shawl.

All the girls admire the shawl of which the pedlar too is very proud. The shy girl retreats, but the others flock closer round him.

4. Then follows a second duet of the pedlar and the shy girl; this time the girl gathers confidence, and she accepts the garment offered her by the pedlar.

The action content is important for the articulation of the movement, as well as for its expression.

4. The pedlar presses the shy girl to take the shawl in her hand. She is thrilled but wonders what the others will think of her if she does. They urge her on, and the pedlar feels he is gaining ground. But she hesitates, torn between her natural shyness and her desire to handle the beautiful fabric.

Then the pedlar, feeling somewhat hurt, offers it to her as a gift.

The girl suddenly realises what is happening, and darts away, embarrassed.

The pedlar is upset, and again approaches the girl, takes her hand and leads her gently into the foreground. Then he wraps the shawl about her shoulders.

Notes to the Second Scene. A suitable musical accompaniment is indispensable in this second scene. The first scene could be played without music, or with an accompaniment of background noise, resembling the noise of the sea. This scene requires gay rhythmic dance music.

The different characters of the three girl buyers will be expressed in their movements. Whatever characters they assume must stand in sharp contrast to the shyness of the young girl who later on gets the shawl. When the whole group rushes towards the pedlar as he produces the golden shawl, surprise and admiration will fix the group movement in an expressive group tension. The first duet is interrupted through this group movement and when the pedlar approaches the shy girl with his gift, everybody watches attentively, and the group surrounding the pedlar opens out to give him free passage to the girl.

When the duet is resumed, the movements of the shy girl become more vivid, corresponding to her eagerness to touch the shawl. There is an increase in her vivacity up to the moment when with a quick, almost brusque movement, the pedlar presents the shawl to her. She now shrinks back, and is nearly fainting when the pedlar leads her into the foreground. Feeling the precious shawl about her shoulders, she is moved to rapture, and expresses her feelings in a short, very light and happy solo-dance, and, at the end faces the pedlar, a grateful and modest expression suffusing her face.

Scene III

Groupings and Movements	Action
1. Two new figures appear. They are middle-aged women whose movements tend to follow straight pathways. The final duet of the second scene is thus broken up. The main group of girls accompany the entrance of the new figures with movements more angular than those they have hitherto used.	1. Two elderly fishermen's wives, one of them the mother of the shy girl, enter, very angry at this merriment. The girl with the shawl hastens away, and hides behind the others. The mothers order the pedlar's wares to be returned to him, at which all the girls laugh. A comic trio develops between the two middle-aged women and the pedlar. The mothers declare they have no money, and cannot buy the wares, much as they would like to please their daughters.
2. The next movement consists of several short solo dances of individual girls who approach the pedlar one after the other, and then return to their starting place. The shy girl is then sent by the women to the pedlar. During her return, there is an emotional break of a sad character.	2. One after the other, the girls return their purchases, one carelessly, another sorrowfully, another rudely throwing hers at the pedlar. The mother eyes her daughter with the shawl. The girl clings to it, but the mother insists that she gives it up. Reluctantly, the girl removes the shawl from her shoulders and returns it to the pedlar. Her disappointment is bitter, and she runs sobbing to her mother. Dismay is discernible on the faces of all.
3. The pedlar performs now a short solo dance carrying and swinging the garment which the shy girl has returned to him. He throws the shawl at her and she catches it. The group of girls, divided into two, encircle one half the pedlar, and the other half the shy girl. A solo dance of the shy girl follows, becoming a duet when the	3. Then the pedlar explains that he intended the shawl as a gift, whereupon the girl's spirits rise again, and she looks about her in wonderment, for she cannot believe that the shawl is not her very own possession.

pedlar joins in. The whole group, and the soloists, stop and look in expectation to the entrance whence the newcomers of the next scene enter.

Her eyes are fixed on the pedlar as he gaily throws the shawl back to her. She catches it with rapturous delight, and swirls round and round in ecstasy. As she sweeps the stage, all the girls and mothers share her enthusiatic delight.

As the first surge of her excitement quietens, she eyes the pedlar more closely. She begins to wonder whether she is worthy of such a gift. She dances shyly round the pedlar, watching intently all the time and, as she retreats from him, he slowly advances towards her and is about to speak to her and her mother when the fishermen come in from their fishing expedition.

Notes to the Third Scene. The choreographic character of the third scene is again a mime-dance. The music chosen must once more be vivid and gay, but the almost grotesque appearance and actions of the mothers must be duly considered. The movements of the mothers, and the group, in this scene differ fundamentally from the movements used in the soft emotional expression at the end of the second scene. The latter might have had a sustained wavy character, but the movements in this third scene will be sharp, angular and energetically stressed. The returning of the gifts to the pedlar by the first three girls constitutes again short solo dances in which the personal character of each of the girls must become clearly expressed. The pedlar is still immobile when the shy young girl returns the shawl. From this moment a new duet between her and the pedlar begins. The other girls join in. Once half of them surround the pedlar and dance with him while the other half dance with the shy girl. When the pedlar and the girl are reunited, and are approaching the mother why is standing dubiously aside from the general gaiety, some of the girls notice, and announce, the return of the fishermen who are not yet on the stage when the third scene ends.

In this, as in all other scenes, it is useful to note, and to consider, the rise and fall of emotional expression. In this scene it could be described as follows.

Entrance of mothers — Mocking excitement.
Returning of the pedlar's wares — Quieter and resentful.

Return of the shawl — Heavy and sorrowful.
Pedlar presenting shawl as gift and duet — Increasing joy and gaiety.
Consent of mother — Solemn, emotional.
Announcement of fishermen's coming — Sudden fright.

Scene IV

Groupings and Movements

1. Several men come in heavy and slow of step. The shy girl dances lightly round and between them.

From here onwards the movements become less dancelike, but the groupings still follow a definite rhythm in space and time.

2. Two middle-aged men walk to the front without paying much attention to the excited and frightened women. Several younger men carrying and handling working implements follow.

3. A trio between the shy girl, one middle-aged woman and her husband (the shy girl's father) is staged, in which the father snatches the garment from his daughter's shoulders.

Action

1. The old fishermen enter. They are morose and disgruntled both on account of the playful women and the pedlar. The girl continues to dance amongst the men, wearing her shawl, but uneasy as she observes their rising tempers.

2. The fathers enter, whereupon the mother quickly takes hold of the shy girl, and leads her behind the other girls.

The fathers are followed by the young fishermen carrying a net and fishing implements. They are dejected and miserable and throw down the net to indicate by that act that they have caught nothing at all.

The mothers, used to an empty net, just glance at it, and, helped by the girls, carry it away.

3. The father sees his daughter wearing the shawl. He is furious. The mother tries to placate him by pointing out how attractive she is, clad in the beautiful garment. But the father snatches it from her, and throws it in the pedlar's face.

The pedlar is dismayed, and tries to explain that he intended the shawl as a gift.

The women advance angrily towards the men. They demand some pleasure in life by way of change from their daily drudgery. The men retaliate, cursing the women for their selfish desire of ease when they themselves risk their lives at sea.

The pedlar tries to ease the situation, and begins all over again to display his wares to some of the girls, and also to two young fishermen who seem more interested in his stock than the others.

4. A duet between the father and the pedlar follows, at the end of which general excitement arises, ending in the pedlar and the father struggling with the whole group of men.

A duet, a fighting dance between the girl's father and the pedlar, follows. The increasing intensity of movement, during which the two adversaries seem to be victorious alternately, is suddenly broken by the final blow which the father aims at the pedlar.

4. One of the fathers bears down on the pedlar, seizes him, and threatens to strike him. He is prevented by the other men who grasp his upraised arm and drag him hither and thither in an attempt to separate him from the pedlar.

Eventually, the pedlar manages to break free, and attacking first one of the fathers and then each one in turn, hurls them all to the ground.

Finally, the pedlar and the girl's father confront each other. A deadly fight between the two ensues, and all the others witness it in horror.

Suddenly the father hits the pedlar such an effective blow that he falls to the ground, all thinking that he is dead.

5. After the downfall of the pedlar the village people stand for a moment motionless. A heavy and slow procession in which the lifeless pedlar is carried away follows.

Everybody except the two middle-aged men exit.

5. Stillness and silence immediately reign. The girl again wearing the shawl rushes to kneel weeping beside the dead pedlar. After a few moments, four of the men lift the pedlar and slowly carry him out, followed by a long procession of

people who have witnessed the catastrophe.

They have no thought at all for the father who remains behind. Only one of the old fishermen stands by him. The mother tries to persuade her husband to return home with her, but in vain. She goes disconsolately away and the two men are left.

Notes to the Fourth Scene. Up to this point in the mime background music similar to that of the first scene is necessary. After the sudden stillness following the death of the pedlar, the scene should proceed without musical accompaniment.

Scene V

Groupings and Movements

1. Duet of the two middle-aged men who have been left behind on the stage. They work agitatedly with the implements brought on to the stage in the previous scene.

Action

1. They stand undecided, and the father begins to fear arrest. He looks in the direction in which the pedlar's lifeless body has been carried away, but there is no sign of anyone.

The two men turn to look at sea, which offers them temporary freedom. They decide to go out for a night catch.

They begin to collect their implements, and light their lamps, for darkness is coming on.

2. Several men, almost the whole group except a few young men, enter marching with heavy and slow steps towards the two middle-aged men.

Mime-movements between the two soloists and the group. The next movement is a common exit of all the men by the same door at which they entered in the previous scene.

2. The other men enter and look suspiciously at what the two are doing. The father approaches, and tells them that they must accompany him to sea. The men are sullen and annoyed with the father for causing this disturbance.

They point to the sky, presaging a violent storm. They protest that it would be madness to embark. The father tells them that they are all in grave danger because of the pedlar's death. After deliberation amongst themselves the men agree, and with heavy steps they go out.

3. Girls rush in at first one by one, later in small groups. Mime-dance follows with swaying rocking movements as if driven by a storm.

The middle-aged women enter, there is a general mood of despair.

In the midst of the swaying main group, and with movements clearly detached from the general background, a short duet between one of the middle-aged women and one of the soloists — her daughter — takes place.

Several girls run in the direction of the men's exit.

The other middle-aged woman hurries out on the other side of the stage.

3. One of the girls sees them as they go out. She pleads with the last of them to stop such folly, but in vain.

She gives the alarm and brings in a second girl.

The mother enters and shouts to the father to come back, but they are already in their boats. One by one, the other girls rush in and call loudly for their sweethearts, but no one answers.

The mother falls sobbing on the shoulder of her daughter. Thunder rolls accompanied by lightening and heavy rain.

Suddenly one of the girls sees a boat capsize. Despair seizes all.

One of the mothers runs to seek men who will volunteer for rescue work. There is anxious suspense.

4. She returns accompanied by two young men who have been left behind: quartet of hurried goodbye with two soloist girls. Exit boys.

Enter the pedlar. The main group is petrified with terror. Short but intensive duet of the pedlar with the shy girl. Exit of the pedlar.

Short solo by the shy girl. The main group continues the swaying dance.

4. The mother returns with two young men who race towards the sea.

They are followed by the pedlar who enters with heavily bandaged head, clutching the shawl. The shy girl rushes to him and tries to dissuade him from the hazardous rescue work, but he quickly gives her the shawl and follows the other men. Sobbing, the girl falls to the ground.

Notes to the Fifth Scene. Background noise indicating the increasing vehemence of the storm is necessary.

The swaying dance of the group of girls should not be too short and must have its own development. Some indications of the mime contents of possible groupings during the swaying dance can be elicited from the inserted duets and the quartet. The main group, however, pays only part attention to these individual outbreaks of passion.

Scene VI

Groupings and Movements

1. The swaying dance of the main group breaks up suddenly.

Several older men are carried lumberingly on to the stage.

2. The pedlar deposits the body in the centre of the stage, supported by one of the middle-aged women, forming a group reminiscent of the Pieta of medieval paintings in which our Lord's body lies across His Mother's knees.

Short duet between the shy girl and her mother (sitting with the father's body across her knees).

3. The group of girls kneel in two groups, right and left to the centre group (Pieta).

The men walk solemnly in a row behind the Pieta group.

4. Short solo of the pedlar. Duet between father and pedlar.

The father with arm outstretched turns towards the pedlar. The ped-

Action

1. There is anxious suspense, and then the first man rescued is brought in by two younger men. His wife and some of the girls cluster round him.

Another man is brought in and is scolded by his wife for his stupidity in embarking in such weather.

2. The mother of the shy girl still waits for her husband. She is about to give up hope when the pedlar enters carrying the father with whom he fought, and by whom he was supposed to be killed, on his shoulders. All wear an apprehensive look.

The mother almost faints. She is supported by one of the fishermen. The pedlar lays the unconscious father at the sitting mother's knees.

3. All think that the father is dead.

The daughter gives the shawl to her mother, who uses it to cover her husband. The women kneel in prayer, while the men draw up, and walk very slowly behind the mother and her husband.

4. Just as the mother is about to cover the husband's face, the pedlar makes a sudden movement and snatches away the shawl. The father

lar advances slowly to meet him, and the people close round upon them.

opens his eyes, and looks round him. There is deep and general amazement.

The father slowly rises, and lifts his arms upwards in thanksgiving.

The men and women draw back from him in wonderment, as though they had witnessed a miracle.

5. The father firmly holds out his right arm, and one by one the men group into a close half-circle, each stretching an arm to the centre of the half-circle, and gripping each other's hands.

5. The father signals to the other fishermen. He asks them if they too will accept the pedlar as their friend. The answer is not in doubt. All place their hands, one by one, on his outstretched arm. At a signal from the father the pedlar advances and is admitted a member of the community.

The women, relieved and thankful, reach out towards the group of men.

6. Duet of father and pedlar. The men's group dissolves. Duet of shy girl and father, who, standing in the centre of the stage, wraps the garment around the shoulders of the shy girl.

Two groups of men and women approach the central group of father and daughter with raised arms as if forming the arcs of a vault above the two central figures. The pedlar takes one step towards the daughter with outstreched arms. The mother of the girl touches his shoulder.

The curtain falls.

6. The pedlar gives the shawl to the father who takes it with seeming reluctance and holds it in front of him. And as he does so the pedlar looks towards the young girl who quietly advances with her mother — the others making way for them.

As the girl arrives in front of her father all raise their arms towards them. The father lifts his hands and brings them to rest with a gesture of blessing on his daughter's shoulders.

Notes to the Sixth Scene. The sounds depicting the storm calm down, and after a very short interval of total stillness when the Pieta group is formed, a solemn harmonious tune of a religious character sets in. The music swells up to the final scene, when the father invests his daughter with the golden shawl.

APPENDIX

Some Fundamental Aspects of the Structure of Effort

Effort in General

The meaning of the word effort* does not only comprise the unusual and exaggerated forms of spending effort, but the very fact of the spending of energy itself. Even the most minute exertion demands some kind of effort. No matter whether the exertion appears to be more bodily or mental, there is always at its origin a process which can be compared to the switching on of an electric current. This primary function is the exclusive privilege of living beings. No inanimate object can make an effort. Out of the storeroom of life energy, which is continuously replenished from birth to death, a spark is detached and used to ignite, as it were, the flow of the mechanism from which the mental-bodily deed results. The inherent capacity of the individual to switch on and ignite the mechanism of inner and outer operations comes to an end when life ceases.

The expression "to make an effort" usually means to spend a considerable amount of mental or physical energy in order to achieve an aim. Nobody is, however, able to exert either his muscular power or his thinking or feeling power in clearly separated ways.

The fact that intense thinking causes bodily fatigue and that physical exhaustion is often due to emotional strain, shows that physical exertion is also contained in an effort of the mind. Similarly, neuro-muscular energy cannot be spent in bodily action without any participation of mental-emotional effort. Therefore, to make an effort involves the whole person.

Through observing and gaining sensitivity to movement it is possible to comprehend the rhythmical complexity which characterises actions.

It has been found that the initiating effort leaves its traces in all man's utterances. Because any utterance is movement, either visible or audible, one can see or hear its distinctive character. The various manifestations of

* *See also* page 21.

169

the initiating effort which can be perceived in human movement may generally be referred to as "efforts", but in order to appreciate the role which each plays within the entirety of movement behaviour it is necessary to discern their individual structures. This is done best by getting first of all the feel of the movement in the body, then interpreting it in terms of effort and finally by writing it down in effort notation.

Mutations

Basic effort actions* are examples of *mutations*. A mutation consists of the change of one basic effort into another. The names of basic efforts exclude each other. It is obvious that a pressing slash or a flicking wring would make nonsense.

Mutated efforts might have two, one or none of the effort elements in common, that means that in mutations, one or more effort elements are changed.

Basic efforts can also be altered without being mutated into other basic efforts. This represents a variation which gives to the effort elements a different *rank* within the same basic effort, i.e. the whole action belongs before and after its variation to the same category of basic efforts.

Ranks

In the following graphs it can be seen that no element is changed but the stress is laid on a different effort element in each variation. Through the concentration of the moving person on only one of the effort elements, a special rank is given to that element, so that it becomes the main one, while the others become secondary in importance. This is represented by a dot (.) next to the main element. There are three variations or derivatives of each basic action.**

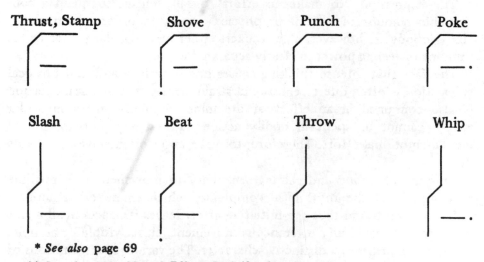

Thrust, Stamp   Shove   Punch   Poke

Slash   Beat   Throw   Whip

* *See also* page 69

** *See also* page 69 and *Effort*, Rudolf Laban and F.C. Lawrence (Macdonald & Evans, 2nd edition 1974). Note, the dot has replaced the previous comma.

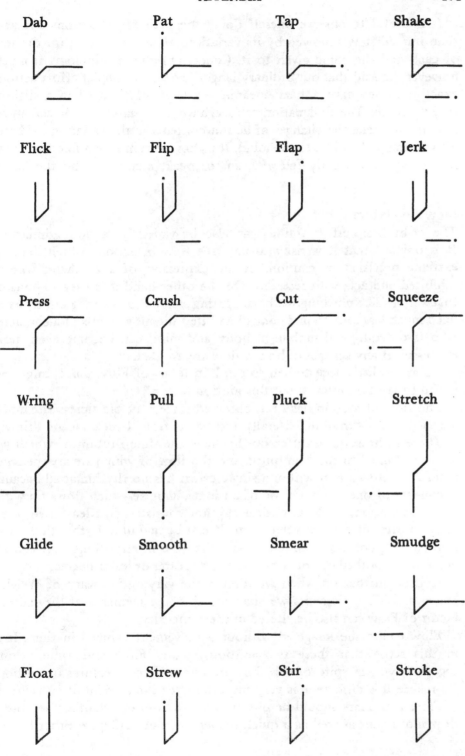

Dab   Pat   Tap   Shake

Flick   Flip   Flap   Jerk

Press   Crush   Cut   Squeeze

Wring   Pull   Pluck   Stretch

Glide   Smooth   Smear   Smudge

Float   Strew   Stir   Stroke

It is useful to observe oneself doing the basic effort actions one at a time immediately followed by its variations. Thus one will learn the feel of each and the name given to it. Concerning the terminology, it must, however, be said that our ordinary language does not depict effort actions exactly; names have a hazy meaning and are often used in a slightly varying sense. The designations above have been carefully chosen in an attempt to make the changes of action contents in the variations of basic efforts comprehensible to the mind. It is, however, indispensible that their real structure is clearly felt and, so to speak, memorised by the body.

Flow Admixture

The eight basic effort actions can also be varied by a flow admixture. It is obvious that it is risky to use free flow in actions which demand extreme precision or caution, or are expressive of a character who is inhibited and acts with restraint. On the other hand it is easy to imagine that actions like plucking feathers, beating a carpet or shaking out a duster need much less caution and control and that the movements of a character who is outgoing and mobile in body and mind can become even more expressive if any vestiges of bound flow are avoided.

We are so little accustomed to think in terms of Flow that it might be useful to say a bit more about this motion factor.*

The flow of motion has been observed since very old times. The much misused word "rhythm" literally translated from Greek means "flow".

There is, however, a difference between the time rhythm of which we are accustomed to think in music and the flow of which we are speaking here. A fluent motion within a single effort has no rhythm at all because it consists of one single release of an inner impulse which flows on without interruption, while a musical rhythm consists of at least two beats with an interruption between them. It can be said of a rhythm that it has a good or poor flow, which means that the interruptions between its beats are smoothed out or controlled to a greater or lesser degree.

A little further on when we discuss the varying intensity of Weight, Space and Time elements we shall see that the elements of the motion factor of Flow can also be altered in their intensity.

Flow cannot be imagined without a movement evolved in time. It is in this sense that these two motion factors Flow and Time belong together, yet, in spite of this they are two different features of motion. To realise this difference is very important for the student of movement.

Flow and Time should be observed apart from one another because a slow movement as well as a quick movement can both have either free or

See also page 75.

bound flow.

At this point it might be mentioned that Weight and Space have also common traits. They can in contrast to the flux and speed of an action be conceived without any perceptible movement in time. However, in spite of their common traits they should be observed as separate features of motion.

Grades of Intensity

A further alteration of effort can be made by differences in the intensity of any one element, which we may refer to as the *grades* of effort elements.

Grades are sometimes adaptations to the needs of an action when dealing with a practical task, and sometimes they arise from an inner mood or disposition. It must, however, be kept in mind that grades are something different from ranks. Ranks show importance, grades show intensity.

For instance, in a cutting pressure it is most important to keep the directness of the movement if a clean straight cut is aimed at. Therefore, the space element "direct" remains the main element in all such cutting pressures while the speed and strength of this action may vary, making the motion factors of Time and Weight secondary (see list of derivatives above).

These secondary elements can be seen to change if the cutting is done with greater or lesser strength or greater or lesser speed, while the directness must remain predominant and unaltered for the action to be efficient. On the other hand if the cutting movement is a symbolic gesture any of the elements might vary in their intensity and even a flow element with a greater or lesser degree of intensity may contribute to the quality of the expression.

We have now seen that any element can be changed, though not always to the advantage of the efficiency of the action. To find the appropriate grades of the intensity of motion is most important in practical actions as well as in dancing, acting and miming.

These grades of any or each of the elements composing an effort might change from an almost entire tuning down to an extreme exaggeration of the intensity of its quality. The most obvious difference is that between a normal and an exaggerated intensity, but moderated or reduced grades can also be observed. Extreme exaggeration of any effort element makes movement almost impossible. This is easily experienced in the rigid cramp resulting from the extremely exaggerated use of strong muscular tension or, conversely, in the sloppiness of an action when the relaxation of this tension is extremely exaggerated.

The Four Grades of Intensity

There are four grades of each effort element, although for practical purposes three may suffice. They are shown in the following way: more than normal "+", less than normal "−"; take for instance:

Dab. Any one, two or all three of its elements can have a different intensity, ranging from reduced, normal, exaggerated to an extreme degree.

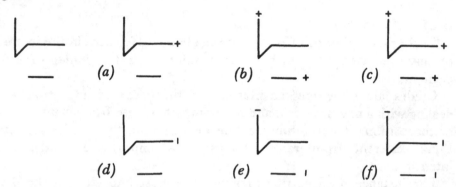

The exaggerations in examples *(a)–(c)* can be detrimental in a practical action yet may bring out poignant expression in symbolic dance movements or in characterisations.

The moderations in examples *(d)–(f)* are indicative of a reduced capacity for the normal intensity of effort but add further shadings to the presentation of human movement behaviour. For instance, our everyday movements frequently tend to involve efforts which contain at least one or two elements of reduced intensity.

If extreme muscular exertion, speed, directness or control is used or their opposites, extreme relaxation, slowness, waviness or release, which demand a highly intense concentration upon one or more of motion factors rather than on the goal to be achieved, then purposeful movement becomes almost impossible. The elements in question are represented by the addition of "++".

Gradations may, of course, also occur in combination with ranks. For instance, a tapping movement may show reduced speed and reduced relaxation:

This means that the predominating directness characteristic of the action is maintained but it has become a little slower and stronger. If a person shows at the same time, say, a strongly suppressed desire to reach and touch something, an exaggerated bound flow will appear as well:

Equally, an effort element which ranks as the most prominant one amongst others may appear in reduced or exaggerated grades. For instance, the important directness in tapping may become exaggerated in the action of a person who is overexacting:

The study of human effort has led to the discovery that three, and only three, alterations in an effort are possible. This discovery is the key to the codification of the enormous number of different shades of effort which can be displayed in movement.

The Three Alterations of an Effort

The three alterations are brought about as we have seen by the gradations, variations and mutations of the effort elements contained in movement.

The practical value of the discovery of the rules of effort alterations consists, firstly, of the possibility of codifying efforts and, secondly, of the distinction between the different characters of motion as observed in nature in general and man's movement behaviour in particular.

As an example it might be mentioned that the movements of mammals are much poorer in their mutations of efforts than the movements of man. Lower animals show in their movements an even more restricted number of variations. The more differentiated actions of which mammals, but particularly man, are capable is due to the greater variety of alterations they are able to produce.

No doubt people whose capacity to grade, vary and mutate their efforts is not well developed are at a great disadvantage as far as the struggle of life is concerned. The pursuit of even the simplest occupation will be hampered and the development of relationships and inner qualities will tend to be colourless and unimaginative. General effort training or the education in the awareness and mastery of the manifold possibilities of effort manifestation in the body thus gains a clear target, namely to develop the individual's capacity to alter his efforts. The insight that alterations follow definite rules through an ascendant scale of complications shows a way in which this can be done gradually and harmoniously.

In reading a person's efforts, the manner in which the differentations of Weight, Time and other motion factors are compounded tends to be little observed and often remains entirely hidden to the unschooled observer.

It is a curious fact that on the whole man is most aware of the Weight factor, while the Space—Time factors recede much further into the region of semi-conscious impressions. Such impressions can be deepened through and for the sake of observation and contemplation. The Flow factor rarely attracts any conscious attention and an extensive mental exertion is needed if one wants to observe it and to understand its role in movement.

It is thus that the gradations and variations of the effort elements concerning weight are the most familiar to us. Alterations of bodily positions and paths in space and differences in speed are easily observed, though their nature is less well understood than that of weight. This gradual decrease of conscious interest for motion factors must be taken into consideration in any attempt to read the efforts shown in a person's movements or to interpret notated effort graphs. The realisation of this is even more important because the difference of interest is based upon the curious habit or capacity of man to use more frequently efforts which are combinations or compounds of Weight and Space elements only. Compounds in which one or both of these factors is lacking seem to be more rare. The Time factor could be said to have a specialised role, namely that of connection or differentation in the building up and dissolving of basic efforts; such fluctuations must be taken into consideration when reading effort graphs.

Limitations in the use of Motion Factors

The first step to the understanding of these fluctuations will be to discuss the general attitudes observable in movement, which we have called that of "indulging in" and that of "fighting against" single motion factors within an effort. These attitudes can be understood as a tendency toward one or the two extreme grades of each motion factor of which the human body is capable. The directing nerve functions as well as the pulley-functions of the muscles and the lever-functions of the bones of the skeleton have limits of capacity in both directions, between the finest attunement and the most powerful resistance to the motion factors. This can easily be understood in connection with weight. A weight exceeding our capacity cannot be lifted and might crush the body if it is too great. Fineness of touch is also limited, because man cannot handle things of infinitesimal smallness of bulk and weight. It is also obvious that the speed of a human movement is extremely limited when compared with, say, that of sound or light. Similarly, we are not able to sustain a movement for that length of time which it takes, say, our body to grow.

More difficult to understand is the limitation of movement as far as its space-filling qualities are concerned. With the utmost flexibility we fill only a very restricted part of a sphere-like space, and what we call a direct movement is never so straight as, say, a piece of string stretched to its full or an undeflected ray of light.

As to flow, it must be noticed that the freest flow we can produce is in reality still slightly bound, as it can eventually be arrested. A really bound flow is near to absolute rest, but absolute rest does not exist in the living body as long as its heart beats and its lungs breathe.

The limitations in the use of motion factors do not, however, impede us from striving for utmost speed, utmost strength, utmost control and utmost directness of our movements or implicitly for all their contrasts. This striving characterises the attitude which we called above "attuning to or resisting" a motion factor.

Another point to which attention must be drawn is the prevalence of Space—Weight combinations in most of our everyday movements. The time used for them may neither be truely assessed as quick, i.e. as having a tendency towards utmost speed, nor can we say that we are using the most deliberate slowness in which movement becomes almost invisible. The neutral speed, a kind of zero point between the limits of extreme quickness and extreme slowness which are available for the human movement apparatus, is the most frequently used speed in our everyday actions. In the recording of efforts we omit the notation of the Time factor when it is neutral.

Flow is also usually neutral in our actions, though like Time, always present in our movements and of very great importance if clearly manifested.

The interesting point is that the four possible combinations of Weight and Space elements are changed through an additional Time element into basic actions.

The mutations occurring through the addition of Time elements to Space—Weight compounds, give us our most familar movements when performing working actions, namely, the eight basic effort actions and their variations.

Between mutated efforts transitions may occur. These transitions follow, when harmonious, definite rules of development according to the activity for which the movement is used.* The limitation to definite forms of transition is one of the most characteristic peculiarities of a movement or effort sequence.

We might say a movement is harmonious if all these alterations and sequences are done in a manner which is in harmony with the human

* *See also* pages 79, 112, 113.

body structure and human movement possibilities. However, movements which are harmonious in themselves do not always guarantee efficiency in work. Some work calls for disharmonious movements producing tensions in the body which do not entirely agree with its best and easiest function. These movements, so far as they cannot be avoided, cause the special exertion which must always be carefully equilibrated by a following relaxation in order to maintain efficiency in everyday life. They are also indicative of disturbances in a person's balance of mental and physical health, but provide the artist with strongly expressive means for his creation of a character or a symbolic dance.

Increases and Decreases of the Grades of Effort Elements

There are occasions when people will be observed using single or compounded effort elements with *increase* of intensity which shows a tendency towards their exaggerated or extreme grades.

In other cases a *decrease* of intensity of an element may be observed which shows a tendency towards its reduced grade or to the changing to the opposite element of the same factor. Increases and decreases belong to the gradation of effort elements and do not change their character. They are marked thus $<$ crescendo, $>$ diminuendo, as from the centre of the graph.

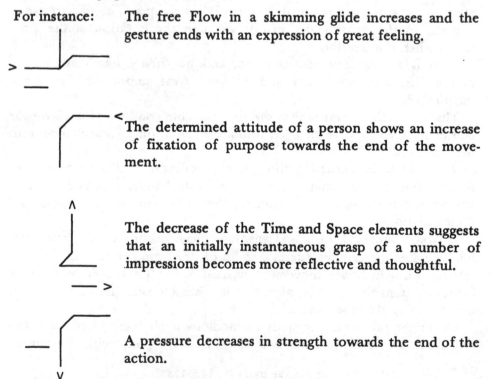

For instance: The free Flow in a skimming glide increases and the gesture ends with an expression of great feeling.

The determined attitude of a person shows an increase of fixation of purpose towards the end of the movement.

The decrease of the Time and Space elements suggests that an initially instantaneous grasp of a number of impressions becomes more reflective and thoughtful.

A pressure decreases in strength towards the end of the action.

The reading of decreasing efforts will be facilitated in considering what happens if an effort element fades entirely away, i.e. if it arrives at the zero point: "/" where no definite grade or character can be observed. Such effort elements tend to change into the opposite element of the same factor, which indicates that a new spark has been initiated, involving the opposite attitude towards it.

This is easy to understand if one realises that any initially firm movement done with steady decreasing strength must arrive at a point where there is no more resistance felt. If the movement continues, there is a probability that the extreme opposite, namely relaxation or weakness, will appear in the action.

Under certain circumstances, depending on the mental make-up of a person, the purely negative weakness might be changed into a positive effort quality of lightness or fine touch.

It is to be noted that increasing effort elements are in no danger of reaching a zero point. The danger here is over-saturation up to an extreme intensity of the same element.

Considering the decreases of the eight elements, if the movement continues, which means that each will be combined with one or more other elements, it can be stated that:

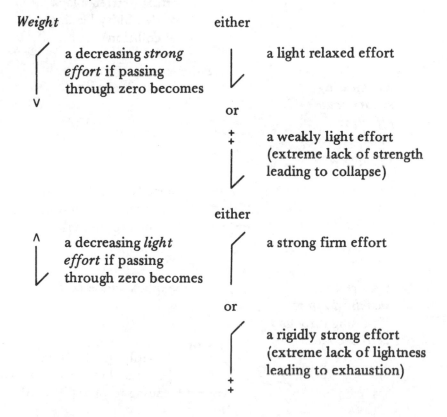

Weight   either

a decreasing *strong*   a light relaxed effort
effort if passing
through zero becomes

                                              or

                                         a weakly light effort
                                         (extreme lack of strength
                                         leading to collapse)

                                             either

a decreasing *light*   a strong firm effort
effort if passing
through zero becomes

                                              or

                                         a rigidly strong effort
                                         (extreme lack of lightness
                                         leading to exhaustion)

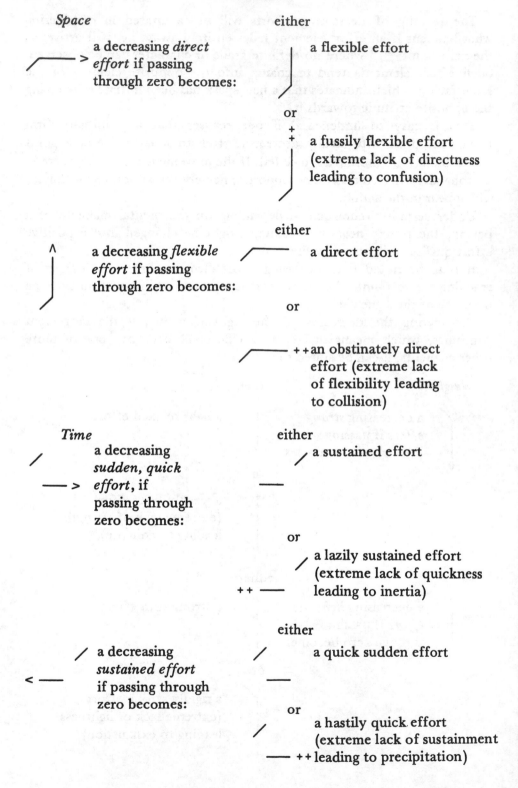

Space   either

/——— > a decreasing *direct*   a flexible effort
        effort if passing
        through zero becomes:

                                                   or
                                                   +
                                                   + a fussily flexible effort
                                                      (extreme lack of directness
                                                      leading to confusion)

                                                   either

^   a decreasing *flexible*   /———   a direct effort
| *effort* if passing
| through zero becomes:
|
                                                   or

                                     /——— ++an obstinately direct
                                              effort (extreme lack
                                              of flexibility leading
                                              to collision)

        Time   either

/   a decreasing   / a sustained effort
        sudden, quick
——— >   *effort*, if   ——
        passing through
        zero becomes:

                                                   or

                                                / a lazily sustained effort
                                                  (extreme lack of quickness
                                             ++ —— leading to inertia)

                                                   either

/   a decreasing   /   a quick sudden effort
        sustained effort
< ——   if passing through   ——
        zero becomes:
                                                   or

                                                / a hastily quick effort
                                                   (extreme lack of sustainment
                                             —— ++ leading to precipitation)

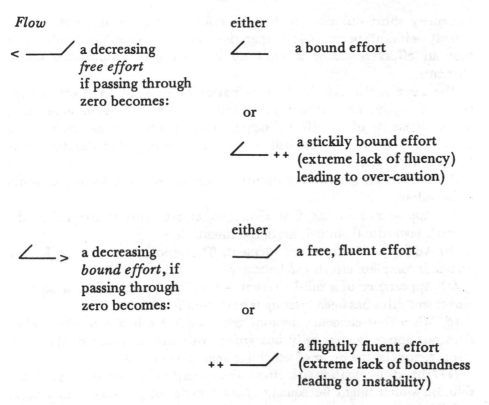

Flow

< ——/ a decreasing
free effort
if passing through
zero becomes:

either

∠— a bound effort

or

∠—— ++ a stickily bound effort
(extreme lack of fluency)
leading to over-caution)

∠—— > a decreasing
bound effort, if
passing through
zero becomes:

either

——/ a free, fluent effort

or

++ ——/ a flightily fluent effort
(extreme lack of boundness
leading to instability)

If all effort elements compounding either an incomplete (combination
of two) or a complete effort (combination of four) or a movement drive
(combination of three) are observed to change their grades simultaneously, this can be shown thus:

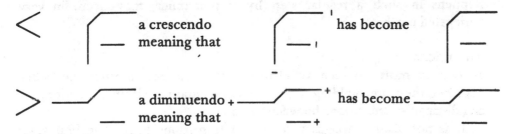

< — a crescendo
— meaning that

— has become

> — a diminuendo +— has become
— meaning that

Building up of an "Effort"

Besides the increases and decreases in the intensity of an effort or its
elements we may get the building up of an effort during motion by the
successive appearance of one effort element after another, and efforts
may disappear in a similar manner. Both appearance and disappearance
show clearly definable principles of orderly succession.

Every effort is initiated by a disposition to act. This is either of extremely short duration so that the effort springs up instantaneously, namely without perceptible preparation, or of perceptible duration, so that an effort sequence is built up by successively appearing effort elements.

The same is the case in the termination of an effort. The effort ends either abruptly, i.e. without perceptible duration of its disappearance or the elements of an effort disappear one by one, so that each of the consecutive stages of the fading out has a perceptible duration thus forming a sequence.

A schematised growth and disintegration of an effort sequence would be as follows.

(*a*) Appearance of the first elemental effort. This is frequently the mover's favourite (habitual) motion element.

(*b*) Addition of the second element. The tendency to use preferably certain incomplete efforts can be stated.

(*c*) Appearance of a third element — whether a basic effort or another movement drive has been built up is here significant.

(*d*) When four elements combine the complete effort is reached. Flow does not appear additionally but grows with the sequence of the other effort elements, the order of which influences its growth.

(*e*) The disintegration of effort shows different possibilities of dissolution which might be equally characteristic for personal habits as (*a*) and (*b*).

(*f*) The last elemental effort to disappear tends particularly to show personal preferences.

Of course, the building up and dissolving of efforts scarcely ever happens in such a regular step by step manner; they occur in very variegated patterns.

Transitions

If two or more consecutive efforts belong to one another, i.e. when together they are building a characteristic quality within an action, we can distinguish *transitions* between them.

It is not always necessary to record transitions fully. The important thing to record is whether an effort appears or disappears with or without perceptible transition from a previous one or into a new one.

Four cases are possible.

(*a*) The effort appears and disappears without perceptible transition.

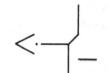

(b) The effort appears with perceptible transition and disappears without perceptible transition.

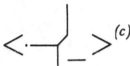

(c) The effort appears and disappears with perceptible transition.

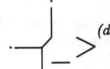

(d) The effort appears without perceptible transition and disappears with perceptible transition.

Each transition would be made between completely different kinds of effort, for example:

and would constitute a short sequence in which the manner of change is notated.

Recovery and Elasticity

Often, we can observe towards the end of a movement or between repeated actions, an effort element changing over into its contrast within the same motion factor. The process seems then to have a kind of recovery or compensatory effect. In the effort graph we can indicate this by a short stroke on the side of the contrast into which the element is changing, for example:

Such changes, when occurring in basic effort actions, are in reality mutations with a transitional incomplete effort between them which is not shown. That means that in one symbol a sequence of efforts is comprehended which indicates an elasticity particularly in repetitive actions.

(a) 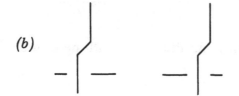 i.e. a thrust developing into or ending in pressing, and its reverse.

(b) 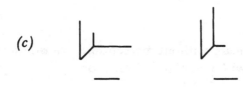 i.e. a slash developing into or ending in wringing, and its reverse.

(c) i.e. a dab developing into or ending in flicking, and its reverse.

(d) i.e. a float developing into or ending in gliding, and its reverse.

(e) 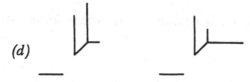 i.e. a press developing into or ending in gliding, and its reverse.

(f) 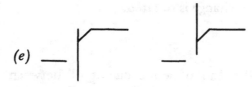 i.e. a wring developing into or ending in pressing, and its reverse.

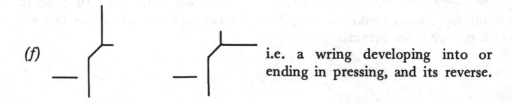

Any single effort graph shows the efforts at a definite time only, during a definite occupation or momentary disposition, or in a definite rest period between actions. Each effort of three or four elements grows perceptibly or imperceptibly from one element to the full compound and each change of effort takes place in a sequence.

In some of the appearances and disappearances of new efforts, a sensible connection between preceding and following efforts can be recognised. That means that any alterations of the elements which may have occurred serve a purpose connected with the aim or object of the motion and its efficient and harmonious performance. Lack of such connections obviously hints at either unintentional incongruities in the mover's mastery of movement, or at an intentional creation of clashes in the relationship between the single parts of a sequence.

Rhythm of an Effort Sequence

The change in which some effort components disappear and new ones appear in a movement phrase constitutes the *rhythm* of the effort sequence.

The mastery of rhythm is an eminent manifestation of bodily intelligence. In observing, for instance, people at games, it will soon be noticed that each player's movements have different rhythms, that means that the parts of their actions are, relatively to one another, of shorter or longer duration, and that some parts are stressed as the main parts of the whole action. The observer of effort will, however, be able to discern yet other relationships besides differences of efforts connected with the Time and Weight factors, namely differences in the use of Space and Flow factors within an effort sequence.

It cannot be too strongly emphasised that interpreting rhythm as a relationship between time durations only, covers but a part of movement observation.

Rhythm can be observed in each individual performance as the order of succession and relationship of efforts within a sequence, as well as in the relationship of several sequences to one another.

Rhythms of effort are almost infinitely variable, but certain principal forms can be distinguished which help the recognition of rhythmical relationships.

An effort can:

(a) stand alone, e.g. between two arrested movements, or motionless periods;

(b) appear and/or disappear, e.g. grow or dissolve bit by bit;

(c) be transmuted through intermediary transitional efforts, in which one, two, three or four elements are progressively changed.

All effort sequences or phrases have a rhythm which may consist of the repetition of any one of the above types of effort or combination of two or more. The beginning and end of a sequence are shown by a vertical double stroke ‖ _ _ _ _ _ _ _ _ _ _ ‖ which also indicates that a pause takes place; this pause may be almost imperceptibly short. Longer pauses can be signalised by a special sign, and pauses can also occur within a longer sequence, breaking it up into shorter phrases. The pause may be just a rest, "∅" during which no particular effort is seen to be active, or one or more elements are maintained thus qualifying the bodily tension during the pause, e.g.   showing it to be flexible and light.

When the same rhythmical phrase occurs again and again, a repetition sign can be inserted and the number of times the movement is performed noted:

Shadow Movements

Shadow movements which precede, accompany, or follow a person's functional efforts, or are otherwise observed in movement behaviour, contain the same modes of attitudes towards the motion factors as movements which have an operational aim with objects.

One of the important forms of shadow moves is the "intentional move". Many intentional moves are preparations for abortive efforts which are finally left unperformed.

Any intention consists of several stages.* The step preceding proper intention is the *attention* given to the object or event upon which intention is focused.

Intention culminates in the *decision* to perform the motion through which the action or the expression is to be perfected. Before the actual deed is effectuated, an anticipation of the final procedure occurs which may be called the *precisional phase*.

Each stage is subjected to a motion factor and its effort elements even if they do not appear as readily visible as fully-fledged operational efforts do.

It will be observed that intentions are sometimes hidden either by an "iron mask" of rigidity, or by movements expressing quite the contrary of the real intention. Such concealment is usually a distinct effort habit and can as such be easily assessed after longer observation.

The hiding of efforts by other efforts creates, in the process of superimposing one effort upon another, a lack of balance in the qualities of the interactions of the various body parts.

* *See also* pages 80, 81, 104, 114.

Similarly, a disturbance of effort balance can be observed in the exaggerated use of pure shadow movements which, in personal behaviour, are discharges of inner tensions. An actor who portrays a highly emotional character may be seen, for instance, to accompany his speech with an array of heterogeneous movement qualities both in his gestures and in his voice. Or, he may be seen to hide his apprehension when meeting a foe by stretching out a hand to him sweetly smiling while his lips are tensely sucked in and his body is rigidly withdrawn.

Since shadow movements originate from the same attitudes to the motion factors as operational movements they are characterised by the same elements of strong or more relaxed tension, of sustained or sudden discharge, of direct or flexible motion, with free or bound flow. They show also grades and ranks of elements exactly like operational moves. The only difference is that they do not intentionally move any outer objects.

In order to distinguish a shadow effort graph from an operational one, the small diagonal line in the centre of the graph which indicates the presence of effort, is crossed by a small stroke.

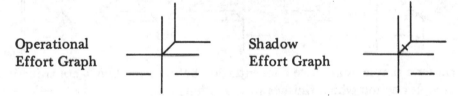

Operational Effort Graph   Shadow Effort Graph

In every day life sequences of shadow movements are rarely repetitive, and they are indeed not memorised or even rarely consciously done. But one can detect an imaginativeness in a person with rich shadow moves and a lack of it in habitually poor shadow moves, the causes of which reach deeply into the sphere of the operations of the mind.

The rhythm of an intentional preparation, together with the following actions, especially if habitual, can reveal certain traits of personality which are characteristic of the mover's manner of acting. One may observe completely contrasting efforts in the inner preparation and the outer action, e.g.:

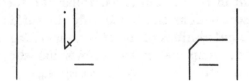

(a) can be interpreted as being jerky in his intentions, but in his operations he shows pushing power;

(b) can be interpreted as being keen and magnanimous in his intentions but at physical work he is cautious and soft.

Or an elaborate preparation may be resolved into a simple action and vice versa, e.g.:

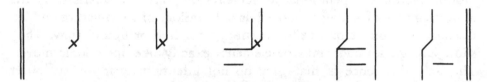

(c) seems to rouse himself gradually as if from a dream and gives himself finally a forceful shake-up before executing a simple slash;

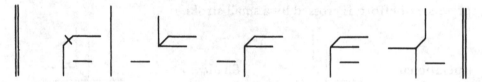

(d) shows a simple immediate concentration before executing a controlled complex action which finishes with a lashing out.

We have previously seen that effort sequences are composed of main efforts, i.e. compounds of three or four elements, of more or less gradual preparations leading up to them, and of their abrupt or step by step dissolutions.

Now we can also incorporate in our observations the effort sequences of shadow moves. The numerous shades of possible preparations, actions, dissolutions arising from the interplay of operational (physical and mental) and intentional efforts constitute rhythmic phrases. Such phrases give the impression of individual units which have interpretable meaning.

Movement rhythms can and do speak their own language to which study of movement in general but in particular of effort, is the best introduction. Verbal explanations can touch only the surface of effort life, if they exceed simple descriptions of the effort happening. An interpretation based on descriptive explanation of efforts has the whole wealth of effort alterations and effort harmony at its disposal.

All the differentiations and their relationships and reciprocal proportions which make up the mobility of effort happenings speak an understandable language in which effort conflicts and their solutions are told.

"Movement has a quality, and this is not its utilitarian or visible aspect, but its feel. One must DO movements just as one has to hear sounds, in order to appreciate their full power and their full meaning."*

* *Effort* by Rudolf Laban and F. C. Lawrence (Macdonald & Evans, 2nd edition 1974). This book and *Personality Assessment through Movement* by Marion North (Macdonald & Evans, 1972) give further insight to the student of effort observation and its practical application.

Index

Index

CPSIA information can be obtained
at www.ICGtesting.com
Printed in the USA
LVHW051953090223
739117LV00005B/380